CIRCUIT ANALY

Theory and Practice, Fifth

LABORATORY MANUAL

Allan H. Robbins
Wilhelm C. Miller

Australia • Brazil • Japan • Korea • Mexico • Singapore • Spain • United Kingdom • United States

Lab Manual to accompany
Circuit Analysis: Theory and Practice, Fifth Edition

Allan H. Robbins and Wilhelm C. Miller

Vice President, Editorial: Dave Garza

Director of Learning Solutions: Sandy Clark

Acquisitions Editor: Stacy Masucci

Managing Editor: Larry Main

Product Manager: Mary Clyne

Editorial Assistant: Kaitlin Murphy

Vice President, Marketing: Jennifer Baker

Marketing Director: Deborah Yarnell

Associate Marketing Manager: Jillian Borden

Senior Production Director: Wendy Troeger

Production Manager: Mark Bernard

Content Project Manager: Barbara LeFleur

Senior Art Director: David Arsenault

Library of Congress Control Number: 2011939565

ISBN-13: 978-1-1332-8102-3

ISBN-10: 1-1332-8102-8

Delmar
5 Maxwell Drive
Clifton Park, NY 12065-2919
USA

Cengage Learning is a leading provider of customized learning solutions with office locations around the globe, including Singapore, the United Kingdom, Australia, Mexico, Brazil, and Japan. Locate your local office at: **international.cengage.com/region**

Cengage Learning products are represented in Canada by Nelson Education, Ltd.

To learn more about Delmar, visit **www.cengage.com/delmar**

Purchase any of our products at your local college store or at our preferred online store **www.cengagebrain.com**

Notice to the Reader

Publisher does not warrant or guarantee any of the products described herein or perform any independent analysis in connection with any of the product information contained herein. Publisher does not assume, and expressly disclaims, any obligation to obtain and include information other than that provided to it by the manufacturer. The reader is expressly warned to consider and adopt all safety precautions that might be indicated by the activities described herein and to avoid all potential hazards. By following the instructions contained herein, the reader willingly assumes all risks in connection with such instructions. The publisher makes no representations or warranties of any kind, including but not limited to, the warranties of fitness for particular purpose or merchantability, nor are any such representations implied with respect to the material set forth herein, and the publisher takes no responsibility with respect to such material. The publisher shall not be liable for any special, consequential, or exemplary damages resulting, in whole or part, from the readers' use of, or reliance upon, this material.

Printed in the United States of America
1 2 3 4 5 6 7 14 13 12

Contents

Preface

This manual provides a set of laboratory exercises that cover the basic concepts of circuit theory. While its 28 experiments are more than can be covered in a normal sequence in an introductory course, it provides flexibility and allows instructors to choose labs to suit their programs. The sequence is also flexible. For example, some instructors may wish to move the introductory oscilloscope lab to an earlier spot—immediately following Lab 8, for instance.

Each lab includes a short overview of key ideas, a set of objectives, a list of equipment and parts, instructions on how to carry out the investigation, appropriate tables for summarizing data, and a set of review questions or problems to test the student's comprehension of the material covered. A mini-tutorial and reference guide to equipment and basic measurement techniques is included at the beginning of the manual to provide the student with a readily accessible source of practical information for use during the lab program. Rounding out the presentation is the tutorial, A Guide to Analog Meters and Measurements, followed by a short section on safety in the laboratory.

NEW TO THIS EDITION

- The oscilloscope material in the Guide to Lab Equipment and Laboratory Measurements has been rewritten to focus on the digital oscilloscope (DSO). (The analog scope description has been retained, but it has been moved to our Web site, where it is available as a downloadable file.)
- The interactive circuit simulation courseware available from our Web site may be used as lab pre-study. Where applicable, choose the desired simulation as instructed on the lab cover page. (For instructions on how to access these simulations, see page vi of this Preface.)
- Labs 29 to 41 from the fourth edition (the *Devices* labs) have been moved to our Web site where they are available as downloads.
- Appendix C *Using Microsoft Excel as a Graphing Tool* has been added to show how Microsoft Excel™ can be used to graph laboratory data.

The equipment needed to run these labs is, for the most part, common equipment of the type found at all colleges. Most experiments, for example, can be performed with just a variable dc power supply, digital multimeters, a signal generator, and a two-channel oscilloscope. A few require additional equipment such as an *LRC* meter (or an impedance bridge), a pair of wattmeters, and a three-phase source. A complete list of equipment and components needed is contained in Appendix B.

While this manual is designed as a companion lab book for the text *Circuit Analysis: Theory and Practice* by Allan H. Robbins and Wilhelm C. Miller, it may be used with any suitable text.

ACCESSING THE ROBBINS AND MILLER WEB SITE

To access the Web site for this book and lab manual, go to www.cengagebrain.com, and in the search box at the top of the page, type in **Robbins and Miller** (or the book title and ISBN). This will take you to the book's supporting Web site.

Allan H. Robbins
Wilhelm C. Miller
March 2012

Acknowledgments

We would like to thank the team at Delmar, Cengage Learning for putting this project together. Specifically, we would like to mention Mary Clyne, Senior Product Manager; Stacy Masucci, Acquisitions Editor; Kaitlin Murphy, Editorial Assistant; Barbara LeFleur, Content Project Manager; David Arsenault, Senior Art Director; Jillian Borden, Associate Marketing Manager; and all their staffs. We also wish to thank the reviewers and users who have provided valuable feedback, as well as the people of Cenveo Publisher Services, for their work in copyediting and preparing the manuscript for publication. We especially thank Dan Moroz, senior instructor in the Electrical Engineering Technology department of Red River College, for his help with some of the new oscilloscope material.

A Guide to Lab Equipment and Laboratory Measurements

Before we get started, let us take a brief look at lab equipment and basic measurement techniques. If you are unfamiliar with test equipment, you should read this section before you begin the lab program and reference it as necessary as you perform each lab.

THE POWER SUPPLY

Your main source of dc in the laboratory will be a variable-voltage, regulated dc power supply. The supply plugs into the ac wall outlet and converts the incoming ac to dc. Such supplies generally include a front panel control for setting the desired voltage and a built-in meter to monitor voltage (and sometimes current). A typical supply is shown in Figure 1. Both single-voltage and multi-voltage units are available. The typical range of output voltage for a laboratory supply is from 0 to 30 volts.

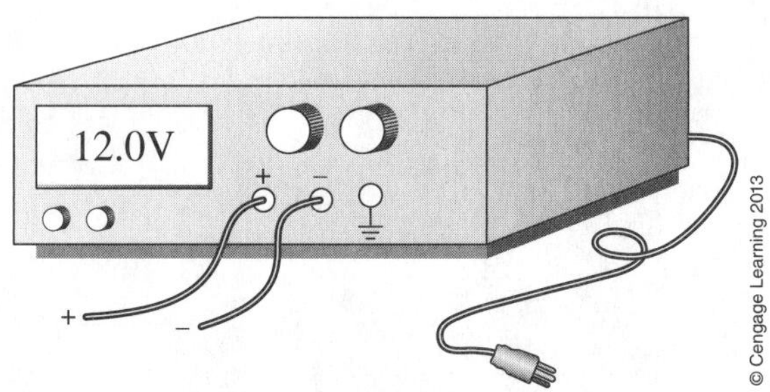

FIGURE 1 Power supply.

Floating Outputs

Most laboratory power supplies have *floating outputs*—that is, outputs where both the positive and negative terminals are isolated from ground. (The source voltage appears between the terminals, and no voltage appears between either terminal and ground.) Such supplies may be operated with one or the other of its terminals jumpered to ground, Figure 2, or both terminals may be left floating. Possible connections are shown in Figure 3.

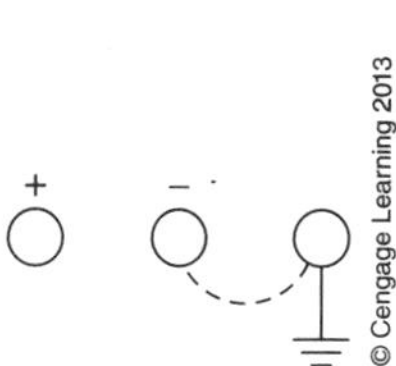

FIGURE 2 Either terminal may be jumpered to ground.

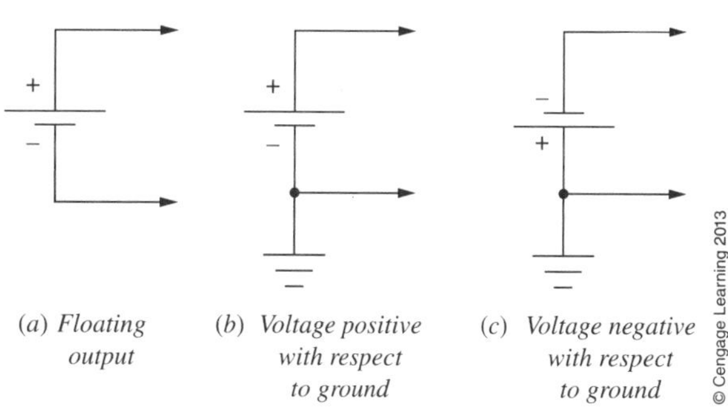

FIGURE 3 Possible connections.

THE DIGITAL MULTIMETER (DMM)

> **Note**
>
> Until relatively recent years, analog multimeters were quite common. They have now been largely superseded by digital multimeters; however, there are situations where analog multimeters are useful. Accordingly, we have included a section entitled *A Guide to Analog Meters and Measurements* later in this manual (following the information on digital meters).

One of the most popular test instruments for basic measurements is the *digital multimeter* (DMM). It combines the functions of a voltmeter, ammeter, and ohmmeter into a single instrument that displays measured results on a numeric output. Figure 4 shows a hand-held DMM. The type of measurement to be made is selected by means of its function selector switch.

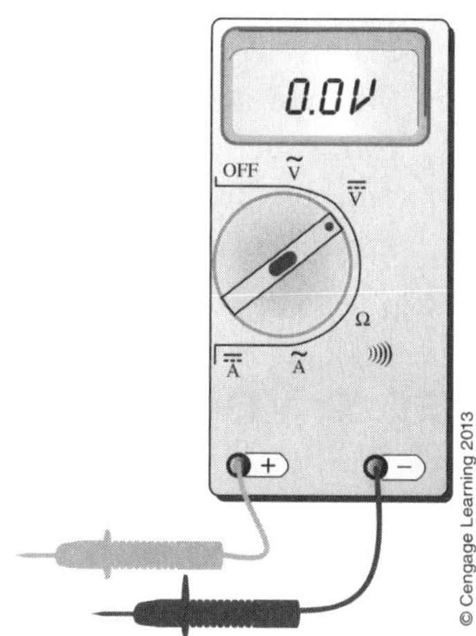

FIGURE 4 A hand-held DMM.

Special Features of the DMM

Two features that most DMMs have are *autoranging* and *autopolarity*. With autoranging, you simply select the desired function (voltage, current, or resistance) then let the DMM automatically determine the correct range. Similarly for polarity—if you connect the instrument for a dc measurement, it automatically determines the polarity (for voltage) or direction (for current) and displays the appropriate sign as part of the measured result.

Manual Range Selection

For non-autoranging DMMs, you must manually select the appropriate range, for example, 250 V full scale, 120 mA full scale, and so on. (Be sure that you do this before you energize the circuit.) If you have no idea of the magnitude of the quantity to be measured, start at the highest range to avoid possible instrument damage and work your way down to lower ranges until you get the best possible reading.

Terminal Connections

As indicated in Figure 4, a multimeter has two main terminals and (although not shown here) several secondary terminals. One main terminal is generally designated *COMMON, COM,* or minus (−). Designations for the other terminals vary. For example, some meters have a common set of terminals for voltage and resistance (which may be designated *VΩ* or plus [+] or some similar designation) plus one or more separate terminals for current. Other meters have a combined voltage/resistance/current input labeled VΩA. Standard lead colors are black and red. By convention, the black lead should be connected to the (−) or COM terminal while the red lead should be connected to the other measurement terminal, that is, (+), VΩ, or A as described later in the measurement examples.

Note

For all labs in this manual, we will assume standard (i.e., non-true rms reading) DMMs with both autopolarity and autoranging.

Average Reading and True rms Reading Instruments

AC meters are calibrated to read rms values. However, most meters (called *average responding*) are designed to measure rms for sine waves only, that is, they are not capable of determining the rms value for nonsinusoidal waveforms such as square waves, triangular waves, superimposed ac and dc, and so on. For these, you need a special DMM called a *true rms* meter. True rms meters are somewhat more expensive than standard meters.

HOW TO MEASURE dc VOLTAGE

Refer to Figure 5: 1) Ensure that leads are plugged correctly into the meter sockets—red lead in the *V* socket and the black lead in the COM socket. 2) Set the function selector to *dc voltage* and select the range if the meter is not autoranging. 3) Connect the probes across the circuit element whose voltage you wish to measure. 4) Read the voltage.

Points to Note with Respect to Figure 5: If the red (+) lead is connected to the positive side of the circuit and the black (–) lead to the negative side, the meter reading will be positive as in (a). Conversely, if the red lead is connected to the negative side of the circuit and the black lead to the positive side, the meter reading will be negative as in (b).

HOW TO MEASURE ac VOLTAGE

The procedure for ac voltage is the same as for dc voltage except that you set the selector dial to *ac voltage*. In this case, however, since an ac meter reads magnitude only, it does not matter which way you connect the meter. This is indicated in Figure 6. As you can see, both connections show the same result.

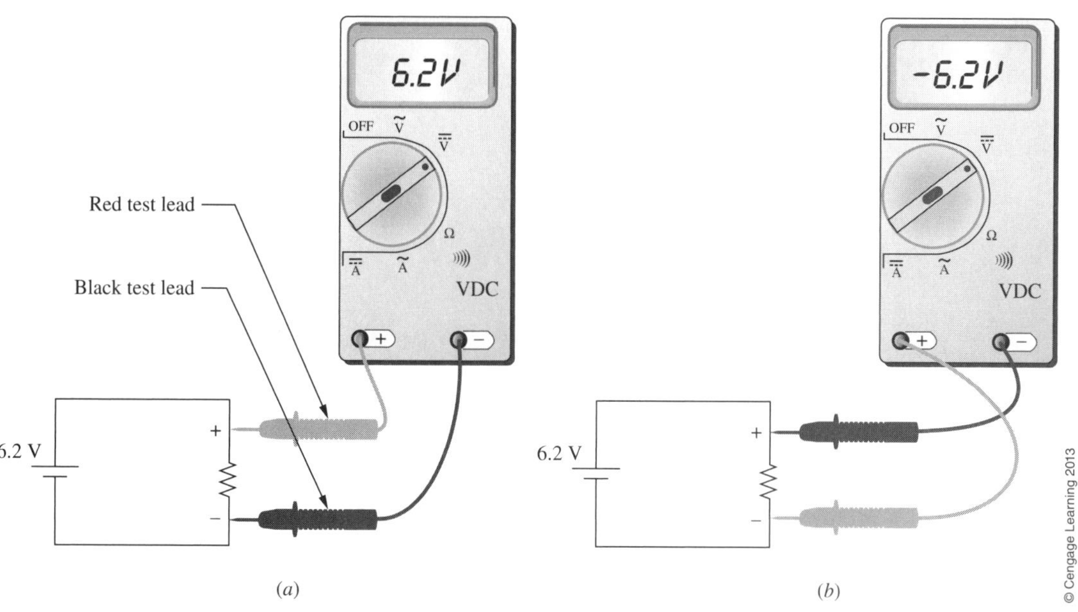

FIGURE 5 Measuring dc voltage. Note that the polarity of (b) is opposite that of (a).

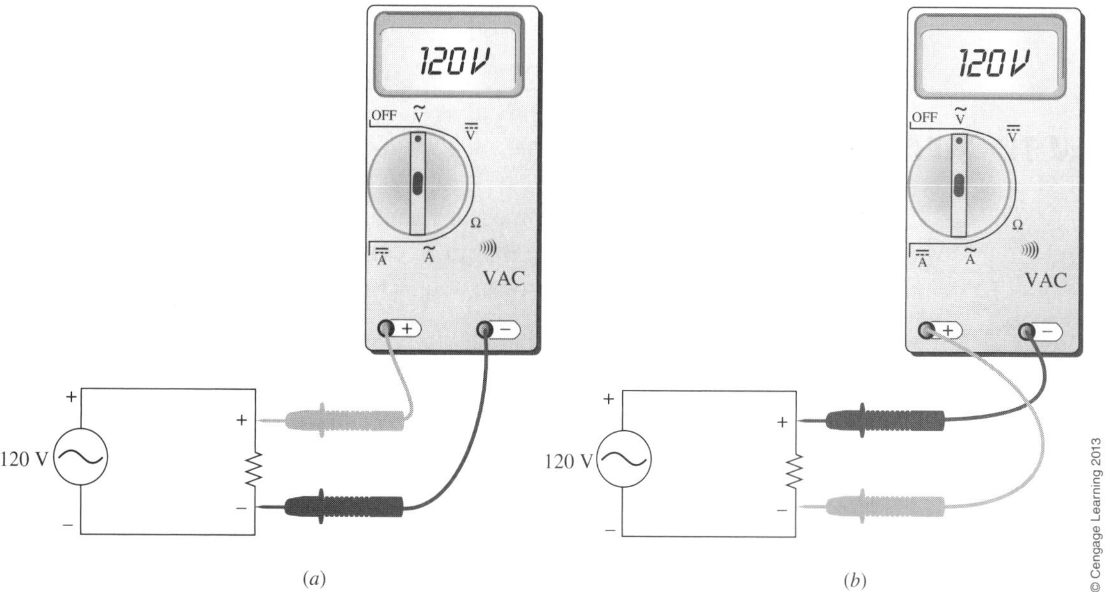

FIGURE 6 Measuring ac voltage. Both readings are the same.

HOW TO MEASURE dc CURRENT

To measure dc current: 1) Turn power off, open the circuit, then connect so that the current you wish to measure passes through the meter as indicated in Figure 7. Ensure that the test lead is plugged into the current jack if the meter has a separate current input. 2) Set the function selector to dc current and select the range if necessary. 3) Energize the circuit and read the current.

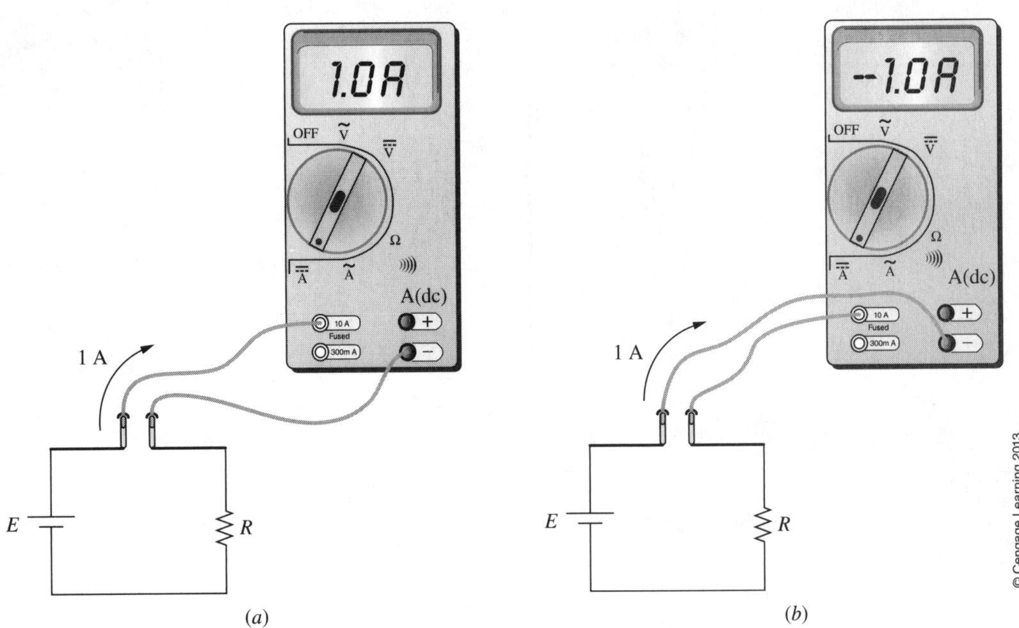

FIGURE 7 Measuring dc current. Be sure to use the current jack.

Points to Note: If the current enters at the A (or VΩA) terminal and exits at COM, the reading will be positive as in (a). If the leads are reversed as in (b), the reading will be negative.

HOW TO MEASURE ac CURRENT

Note: Not all multimeters measure ac current. For those that do, proceed as for the dc case, except set the dial to ac current instead of dc. [Since an ac meter reads magnitude only, you can connect as in Figure 8(a) or (b).]

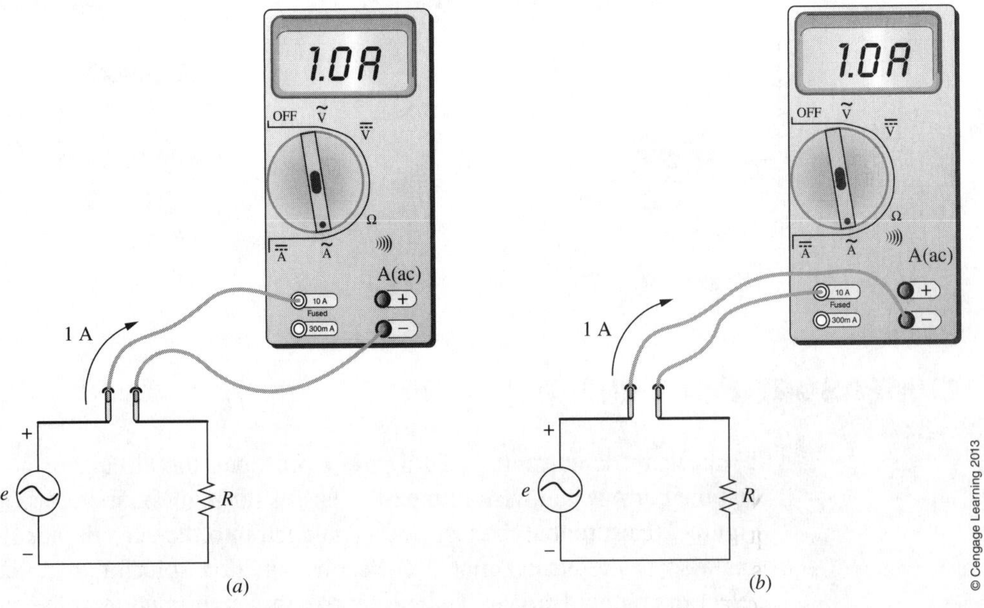

FIGURE 8 Measuring ac current. Be sure to use the current jack.

HOW TO MEASURE RESISTANCE

Figure 9 shows a DMM connected to measure the resistance of an isolated resistor. 1) Set the dial to resistance. 2) Connect the probe across the component whose resistance you wish to measure. 3) Read the measured value. Be sure to note the unit, Ω, kΩ, and so on.

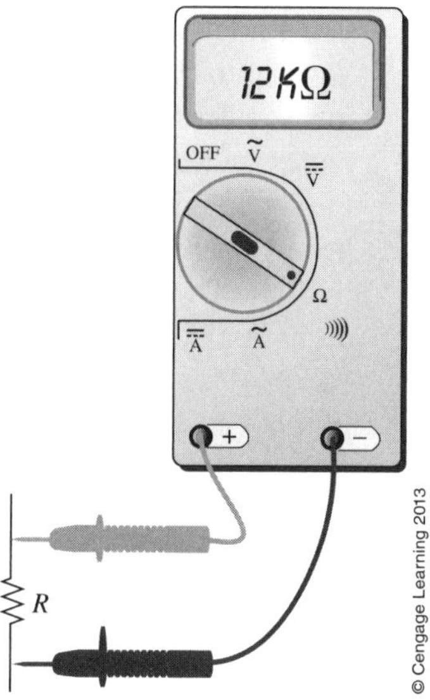

FIGURE 9 Measuring resistance.

Additional Points to Note

1. When connecting the probes across the component to be measured, be careful not to touch both probe pins of the meter, since your body's resistance may introduce additional error. When measuring resistance, the red and black leads of the ohmmeter can usually be interchanged without affecting the reading. The exception is diodes and most active components such as transistors and integrated circuits (ICs).
2. When measuring the resistance of a resistor that is part of a circuit, disconnect the power from the circuit; otherwise the ohmmeter may be damaged. In addition, it will be necessary to isolate the resistor from the rest of the circuit. You can do this by disconnecting at least one terminal of the component from the circuit.

METER ERRORS

No meter can be guaranteed to be 100% accurate. Thus, when you measure a quantity, there will always be some uncertainty due to the meter itself. (This is similar to the situation with bathroom scales. If your scale is not perfectly accurate, it may indicate a few pounds more or less than your actual weight.)

Meter Accuracy Specifications

The *accuracy specification* for a meter defines the maximum error that it may have—that is, the guaranteed maximum value by which the measured value may deviate from the true value.

The basic dc voltage accuracy specifications for bench/portable and handheld DMMs ranges from about 0.05% to 0.5%, depending on make and model. This specification is independent of reading—thus, a DMM with a 0.5% error specification will have no more than 0.5% error no matter what value it is reading. For example, if its displayed value is 100.0 volts, it could have up to 0.5% × 100 volts = 0.5 V error, meaning that the true value of the measured voltage can be anywhere between 99.5 and 100.5 volts. The same meter indicating 16.00 volts still has a possible 0.5% error in its reading (which in this case is 0.5% × 16 volts = 0.08 V) and thus, its true value could be anywhere between 15.92 and 16.08 volts.

Some DMMs also have an additional uncertainty in their least significant digit. For example, a certain popular low-cost, hand-held DMM has a specification of ±(0.5%+1 digit), which means that its maximum error at any point will not exceed one-half a percent of its indicated reading plus one least significant digit. Thus, if this meter is displaying 100.0 volts, it could have an error of half a volt (as noted above) plus 1 digit. This means that the true value of the measured voltage lies somewhere between 99.4 and 100.6 volts.

CURRENT MEASUREMENT ACCURACY VERSUS VOLTAGE MEASUREMENT ACCURACY

Depending on the meter you use, the accuracy of your current measurements may be poorer than the accuracy of your voltage measurements. For example, for the DMM mentioned previously, the accuracy specification drops to ± (1.5% of reading + 2 digits) for current. On the other hand, some meters have the same accuracy specification for current as for voltage. Check your meter manual.

A FINAL NOTE ON METER ERROR SPECIFICATIONS

Accuracy specifications define a meter's worst-case error, and most meters will perform better than their published accuracy specifications. Still, the published figure is the only guarantee that you get. Also note that accuracy specifications for ac measurements may be different than for dc. Always check your meter manual.

FREQUENCY CONSIDERATIONS

The frequency range of a DMM is an important consideration for ac measurements. For example, some DMMs can measure up to only about 1 kHz, while others can measure up to 100 kHz or higher. (If you need to make measurements at higher frequencies, you may have to use an oscilloscope, considered later.)

SCHEMATIC SYMBOLS FOR METERS

So far, we have shown meters in pictorial form so that you can see how the meters are physically connected into a circuit and how the function selector switch is used. In practice, this is cumbersome, and meters are usually represented by simple schematic symbols as in Figure 10. (In most labs, we show only the schematic. However, in the first few labs, we sometimes show both.)

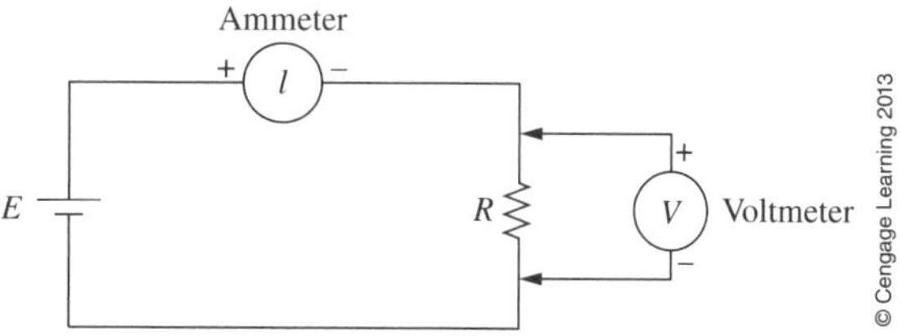

FIGURE 10 Representing meters using schematic symbols.

WORDS OF CAUTION

1. Never connect an ammeter across a voltage source. An ammeter is virtually a short circuit, and damage to the meter or the source may occur.
2. For safety, always plug the probe leads into the meter sockets before connecting the probe tips to the circuit.
3. When using a meter with manual range selection, set the meter to its highest range if you do not know the approximate value of the quantity to be measured, then switch to lower ranges until you get to a suitable range.

THE FUNCTION GENERATOR

A *function generator*, Figure 11, is a variable-frequency, multi-waveform source that produces a variety of waveforms such as sine, square, triangular, pulse, and so on. It will have a set of push buttons or a selector switch for selecting the desired waveform, range selector push buttons or a selector switch for setting the frequency range, a variable control for adjusting the

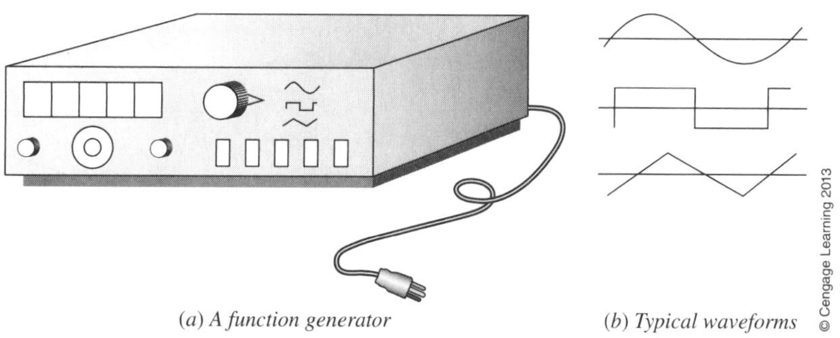

(*a*) *A function generator* (*b*) *Typical waveforms*

FIGURE 11 A function generator and some of its waveforms.

frequency within the selected range, an amplitude control for adjusting the output voltage, and a digital readout (or an analog scale on the dial) for displaying the chosen frequency. A *dc offset* control is usually included so that you can add a positive or negative dc voltage component to the waveform. Depending on the make and model, other features may also be included. The usual frequency range of a function generator is from a few Hz to a few MHz.

THE OSCILLOSCOPE—INTRODUCTION

The *oscilloscope,* Figure 12 (also known as a scope) is an electronic test and measurement instrument. Its main purpose is to capture and display waveforms on a screen as in Figure 13. Since the oscilloscope shows you exactly what the electronic signals of interest look like, it is key to studying and understanding time-varying waveforms. You can use it, for example, to measure the frequency and period of repetitive waveforms, to determine the rise and fall times of pulses, to find the phase difference between sinusoidal signals, and to help troubleshoot electronic equipment. By viewing signals in this manner, you can determine whether a component or a subsystem is behaving properly or is perhaps on the verge of failure.

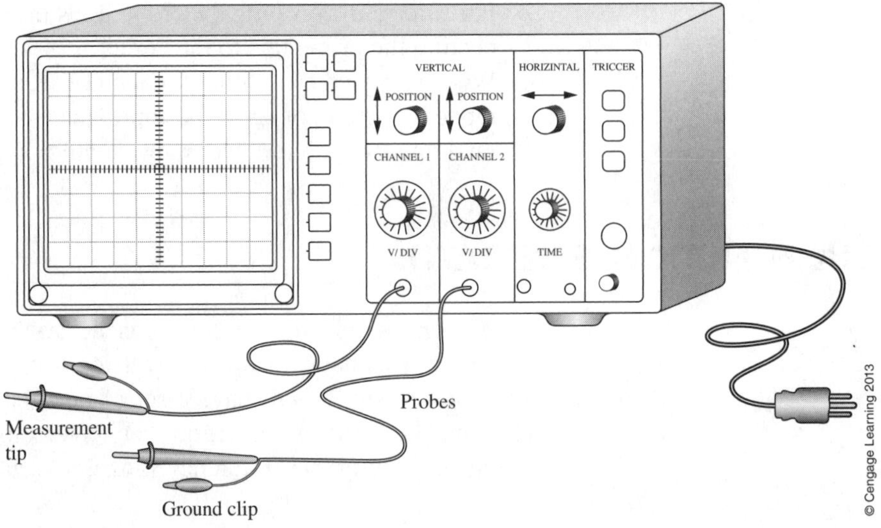

FIGURE 12 A typical dual-channel oscilloscope. Oscilloscopes are also referred to as scopes.

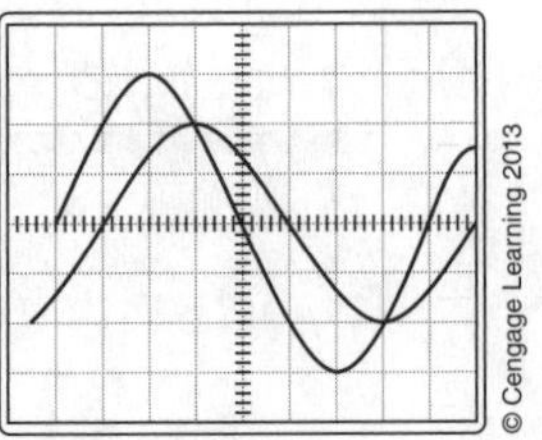

FIGURE 13 Waveforms on the screen.

Since an oscilloscope is a voltage measuring device, you cannot measure anything but voltage with it. If you want to measure other variables (such as current, temperature, or liquid flow rate), you need a sensor to convert the original signal into an equivalent voltage. Although signal conversion is not a part of this discussion, suffice it to say that such sensors are relatively common—for example, a microphone is a sensor that converts sound to an equivalent voltage that the oscilloscope is capable of displaying.

Since an oscilloscope is a rather complex instrument, we look at it in several stages. In this section, we look at it from a conceptual viewpoint; in the lab section, we learn how to use it to make voltage, current, phase angle, and other measurements.

Digital Storage Oscilloscopes (DSOs)

Up until a few years ago, the majority of oscilloscopes were analog and used a CRT (cathode ray tube) for display. However, digital scopes have now largely superseded analog scopes, so in this manual, we concentrate on them. (An analog scope version of this write-up is posted on our Web site. However, it should be noted that, while analog scopes are still used by many people, they are seldom sold anymore.) One of the great advantages of digital oscilloscopes is their ability to store waveforms for later analysis. Because they can store data, digital oscilloscopes are sometimes referred to as digital storage oscilloscopes, but here we simply call them digital oscilloscopes. (A variation of the digital oscilloscope is the digital phosphor scope, but we do not consider it here.) Digital scopes do not use a CRT display—instead, they use a flat panel LCD display. Consequently, the physical size of the oscilloscope has been greatly reduced.

Basic Digital Oscilloscope Block Diagram

Figure 14 is a simplified block diagram of a DSO. (While all oscilloscopes these days are multichannel instruments, we show only one channel for illustration.) The diagram is quite simplified and does not show all control paths required to make the system function. However, enough is shown to present the general idea of what is happening. Although not shown explicitly, the system is controlled by one or more embedded microprocessors.

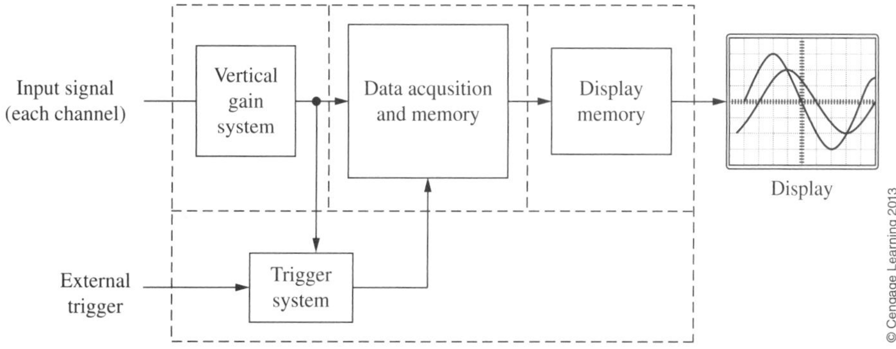

FIGURE 14 Simplified block diagram of a digital storage oscilloscope (DSO).

Stated very simply, the input/data acquisition section, aided by the triggering circuit, samples the applied signal and converts it into digital "samples" which are stored in the *data acquisition memory*. The data is then pulled from this memory, processed, then sent to the display. (DSOs generally have some math capability and are able to automatically compute frequency and period and peak values, add and subtract waveforms, plus perform various other mathematical manipulations. However, the processing capability of a DSO depends on its price, and in this lab manual, we consider only rather basic capabilities.)

Using the Oscilloscope

Although an oscilloscope is a complex instrument, with a bit of experience, you should find that its operation (at least at the basic level), is reasonably easy and straightforward. Typically, you operate a scope using its front panel knobs and buttons. From the user's perspective, an oscilloscope consists of several different systems, each of which you interact with in some way when you make a measurement. Let us briefly consider each in turn (with more details to be added later as you go along).

The Input System

The number of input channels on an oscilloscope is typically two or four. Channels are typically identified as *Ch1, Ch2,* (or perhaps *ChA, ChB*), and so on. Channels can be set for dc coupling, ac coupling, or 0 V (to be described in detail later).

Let us consider a typical channel. Because input signals may be large or small, the *vertical gain* block incorporates both an input *attenuator* and an *amplifier*. A front panel control (calibrated in volts per division) permits you to set the scale factor to suit the measurement—that is, if the signal is too small, it is amplified; if it is too large, it is attenuated. (Some of the attenuation may be in the scope probe itself; this is described in the "Probes" section that follows.) The signal is then sent to the *Data acquisition subsystem,* which contains an *analog-to-digital converter* (*ADC*). The ADC samples the applied signal and converts it into digital "samples" which are stored in the *acquisition memory*. The input is also routed to the *trigger system* (described below) which, in conjunction with the *horizontal system* (described below), provides information to the microprocessors to help control the sampling.

The Display System

The display screen of a digital oscilloscope generally consists of a TFT (thin-film transistor) LCD panel. The display screen is usually 10 cm (about 4 inches) across and is ruled with a grid at 1-cm intervals with graduated markings between grid lines. (These grid lines, called graticules, consist of major and minor scale divisions.) Waveforms are displayed on this grid, and values may be read from its scale as illustrated in Figure 13. (Although with modern oscilloscopes, measurements can be made more accurately using cursors or by the automated measuring system of the oscilloscope itself, it is sometimes helpful to go back to basics to develop a better understanding of what the scope is actually doing.) By selecting appropriate vertical and horizontal scale factors, you can, for instance, determine the amplitudes of the waveforms, their period, and the phase displacement between them.

To display a waveform, the scope essentially graphs it on the screen in the same sense that you can draw a graph on a piece of paper—you can move your pencil vertically to the desired Y-axis position and horizontally to the desired X-axis position, then place a dot. (If you use enough dots placed closely enough together, you get what appears to be a continuous curve.) In a conceptually similar process, the scope electronics do the same thing. For this reason, the scope needs a vertical display/control system and a horizontal display/control system. Let us consider each of these in turn.

The Vertical Display/Control System

The primary vertical display control is the vertical gain control; you use it to set the volts per division (V/Div) for the channel. To illustrate, consider Figure 13. Here, if you have set the control at 5 V/Div and each major graticule line represents 5 volts, the amplitudes of the two waveforms illustrated are 10 V and 15 V, respectively.

A switch on the input line allows you to select *ac coupled* or *dc coupled.* The switch also includes a 0 V input (also called a GND or ground). When set to *dc*, the input signal is applied directly to the vertical deflection system, permitting the entire signal (ac and any dc component present) to be viewed on the screen; when set to ac, the input signal passes through a capacitor which blocks all dc and results in only the ac signal being displayed. The GND position permits you to establish a 0-V baseline by internally applying zero volts to the vertical deflection system.

The scope also has a vertical position control. With this, you can change the position of your waveform by moving it up or down. You can use this feature, for example, to compare data by aligning waves from different channels, one on top of the other.

The Horizontal Display/Control System

As noted earlier, an oscilloscope graphs waveforms of voltage versus time. A horizontal time base control permits you to set the horizontal time scale. It is calibrated in units of time per major graticule division—for example, sec/Div or ms/Div. This control needs to be set according to the frequency of your input waveform; for example, if you are measuring a high-frequency waveform, you need to set the horizontal sweep at a high rate.

Your scope will also have a vertical horizontal position control. With this, you can change the position of your waveform by moving it to the left or to the right on your screen.

Trigger Controls

A scope's trigger circuit determines when the oscilloscope starts to acquire data. It is your (the user's) job to set up the triggering controls properly. For example, suppose you want to capture a trace as illustrated in Figure 15(a).

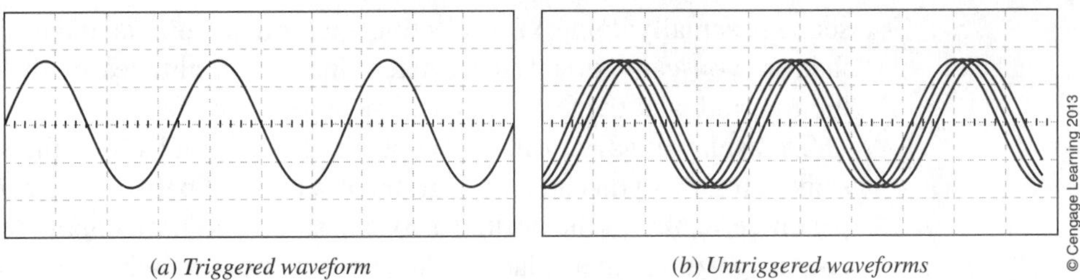

(*a*) *Triggered waveform* (*b*) *Untriggered waveforms*

FIGURE 15 The importance of proper triggering.

When triggering is set up properly, you get a stable trace on your screen; when set up improperly, the trace may drift across your screen, or you may get multiple "ghost" images as in (b), or the screen may simply remain blank.

Triggering options include specifying which channel to trigger from, whether to trigger on a positive or negative slope of the waveform, and at what level the signal should be to start capture and display. (We investigate these options in the lab section.) Other triggering options may also be available on your scope—check the scope manual or ask for assistance.

Frequency Range

Oscilloscopes have an extremely broad frequency range; even low-cost models can measure signals from dc to 25 MHz. Top-of-the-line models can measure signals up into the GHz range.

Probes

Probes (see Figure 12) are an integral part of a scope's measurement system. A great majority of probes used with scopes are passive. Passive probes are usually either 1× or 10× (times one or times ten). A 1× probe has no attenuation, and thus the full voltage at its tip is applied to the scope. A 10× probe, on the other hand, has a built-in ten-to-one voltage divider, and thus, only one-tenth of the measured voltage is applied to the scope. With a 1× probe, the input resistance of the scope is typically 1 megohm; with a 10× probe, it is typically 10 megohms. Thus, the 10× probe results in less loading of high-impedance circuits. Each probe has a built-in *compensation circuit* which must be "tuned" to match the channel to which it is connected. Depending on the manufacturer, this is done by twisting the barrel of the probe or by adjusting a compensator screw on the probe. To tune the compensation circuit, touch the probe to the calibration test point on the front panel of the scope and adjust until the waveform seen on the screen is a flat-topped square wave. Once the probe is tuned, it will not introduce distortions into the measurements that you are making.

Measurements from Waveforms

Graticule Measurement Method: The most basic (but least accurate) way to make measurements is to use the screen graticules. For example, as noted before, if the vertical gain control is set to 5 V/Div, the waveforms of Figure 13 have amplitudes of 10 V and 15 V, respectively. While this is not a precision measurement, it provides a quick, visual estimate.

Cursor Method: A more accurate method is to use the scope's cursors. Using scope controls, position the cursors as applicable and read values—for example, voltage amplitude or time from the waveform.

Automatic Method: This is the most accurate method because the scope makes measurements directly from acquired data. In this case, the scope calculates results and displays them numerically on your screen. A typically automatic measurement is input waveform frequency.

Final Comments

Although most digital oscilloscopes have the previously described basic features in common, there are differences in detail that you will have to learn from your instructor or from the scope's manual. Much of this you will learn as you perform the labs in this manual. In addition, many high-end scopes have very sophisticated features, but they are beyond the realm of this introductory course and will be encountered in your later, more advanced studies.

References: Tektronix TDS 1000/2000-Series Digital Oscilloscope User Manual
Agilent Application Note 1606

Both references are available on the Internet.

A Guide to Analog Meters and Measurements

For general-purpose measurement and testing, digital multimeters are now much more widely used than analog multimeters. However, analog meters are more suitable than digital meters in some applications, since analog instruments can display a changing reading in real time, and are thus able to track trends, identify peak and null values, and so on. (In this type of application, digital instruments are basically useless.)

Analog meters are found in practice in the form of multimeters as well as panel meters. Although we concentrate on multimeters here, the principles described are applicable to all analog meters.

VOMs

Analog multimeters are frequently referred to as VOMs, since they are multipurpose instruments that can be used to measure volts, ohms, or milliamps. Figure 16 shows a typical meter. As can be seen, it uses a needle pointer and a set of scales to display measured values. A function selector switch provides the means to select the quantity to be measured and to specify the scale to be read. Note that the range selector switch must be manually operated; VOMs do not have autoranging capabilities as DMMs have. You must set the selector switch to the appropriate range—for example, 250 V full scale, 120 mA full scale—before you connect the meter. (This is discussed in more detail later.)

Terminal Connections

Like a DMM, a VOM has two main terminals and one or more secondary terminals. One main terminal is generally designated *COMMON, COM,* or minus (−). Designations for the other terminals vary. For example, some meters have a common set of terminals for voltage and resistance [which may be designated VΩ or plus (+) or some similar designation] plus one or more separate terminals for current. Other meters have a combined voltage/resistance/current input labeled V*ΩA*. Standard test lead colors are black and red; the black lead connects to the (−) or COM terminal while the red lead connects to the other main terminal, (+) or VΩ or the appropriate secondary terminal, typically A or mA.

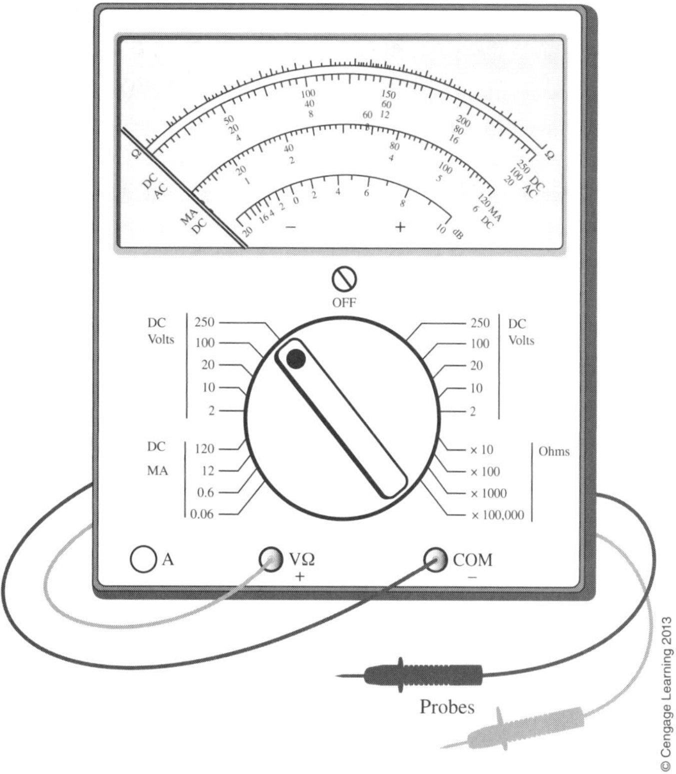

FIGURE 16 An analog VOM.

HOW TO MEASURE VOLTAGE, CURRENT, AND RESISTANCE WITH A VOM

The same basic principles apply to measuring voltage, current, and resistance when using a VOM as when using a digital DMM, so consult the diagrams and discussions in the DMM sections earlier. However, here are some important differences to note:

- Analog VOMs are not autoranging: you must select the range yourself. If you do not know the approximate value of the quantity to be measured, set the meter to the highest range possible (before connecting the meter), then switch to successively lower ranges until you get a good reading.
- Analog VOMs do not have autopolarity. The pointer rotates upscale, but if you connect a meter backward (for dc measurements), the pointer attempts to pivot downscale, but is unable to make a measurement because it comes up against the end stop. In this case, you must reverse the probe connections at the point of measurement.
- For ac measurements, polarity is immaterial and the previous point does not apply.
- VOMs have lower impedance than DMMs; hence, they load circuits more than do digital meters.
- Compared to typical digital DMMs, analog multimeters have higher frequency responses, typically up to about 100 kHz.

- While analog VOMs are calibrated to read rms for ac sine waves, they are generally average-responding meters. This means that they cannot measure the rms value for any other waveform.

INTERPRETING ANALOG METER SCALES

Analog multimeters have multiple scales, and values are more difficult to interpret than for digital instruments. To interpret a result, you must note both the scale reading and the range to which the function selector switch is set. For example, in Figure 17, the dial is set to the 250-V range and you must therefore read the voltage on the 250-V scale. Here, V = 150 V.

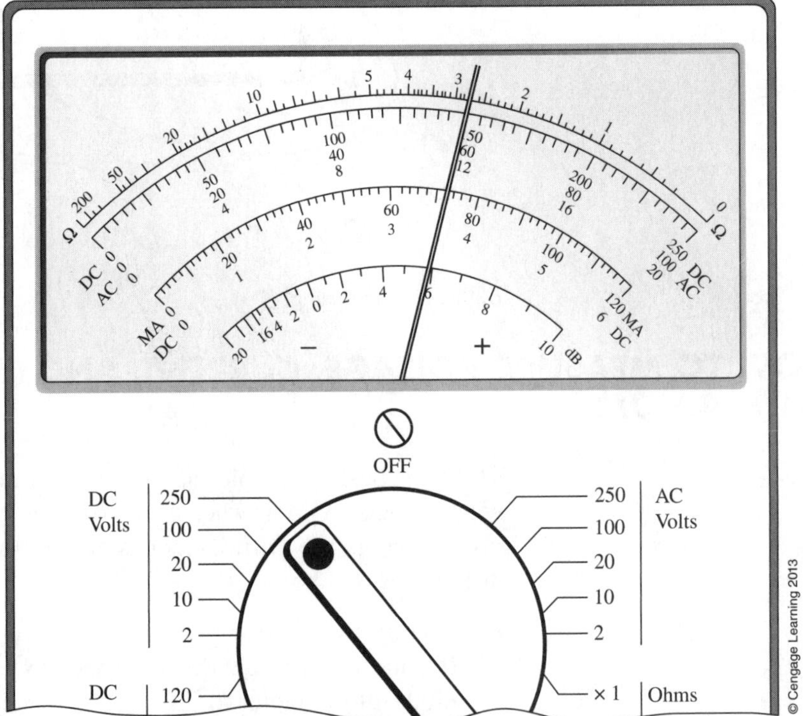

FIGURE 17 Reading an analog scale. V = 150 V.

Note that analog scales are divided into intervals (called *primary divisions)* with unmarked graduations (called *secondary divisions)* between. For example, on the 250-V range of Figure 17, each primary division represents 50 V. By counting the number of secondary divisions between 50-V markings, you can see that each represents an increment of 5 volts. Reading a value is straightforward if the needle is over a primary division as indicated in Figure 17. Reading a voltage is also straightforward if the needle rests over a secondary division as in Figure 18(a). By counting the number of 5-V increments, you can see that the meter indicates 160 V. However, if the needle rests partway

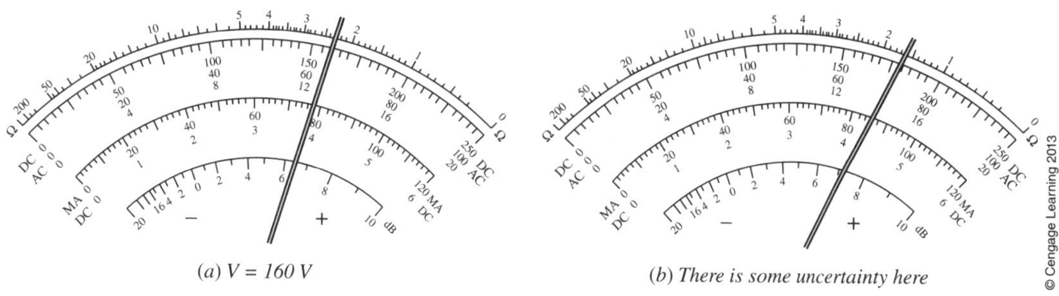

(a) V = 160 V

(b) There is some uncertainty here

© Cengage Learning 2013

FIGURE 18 Interpreting meter scales. The meter here is set to the 250-V range.

between graduations, you must estimate the reading. For example, the reading shown in Figure 18(b) might be interpreted by one person as 183 V but by another person as 184 V. The estimated digit will have some uncertainty due to your inability to read it exactly.

Range Selection

As noted, some scales are used by more than one range. For example, in Figure 17 the 100-V scale is used by both the 100-V and the 10-V ranges. When the selector switch is set to 100 V, full-scale voltage is 100 volts and the scale is read directly, with each secondary division representing 2 volts. However, when it is set to 10 V, full-scale voltage is 10 volts and readings must be divided by 10. In this case, secondary divisions represent 0.2 volts. Thus, for Figure 17, if the range selector were set to 100 V, the reading would be 60 V, while if it were set to 10 V, the reading would be 6 volts.

METER ERRORS

No meter can be guaranteed to be 100% accurate. Thus, when you measure a quantity, there will always be some uncertainty due to the meter itself. In general, DMMs are more accurate than analog VOMs.

Meter Accuracy Specifications

The *accuracy specification* for a meter defines the maximum error that it may have—that is, the guaranteed maximum value by which the measured value may deviate from the true value.

The accuracy of a VOM is defined in terms of its full-scale value. Thus, a VOM with an accuracy specification of 2% measuring a voltage on a 50-V scale may have as much as 1 volt error anywhere on that scale (since 2% of 50 V is 1 V). While this 1-volt error represents 2% at full scale, it represents 4% at half scale and 10% at the 10-V point. Thus, for a measured value of 50 volts, the true value may lie anywhere between 49 and 51 volts, but for a measured value of 10 volts, the true value may be anywhere between 9 and 11 volts. Accuracy can be improved by changing to a range where the reading is closer to full scale. For example, a 2% error on the 10-V range represents a possible error of 0.2 volts;

thus, if you measure 10 volts on the 10-V range, the true value lies between 9.8 and 10.2 volts, a considerable improvement over the same reading taken on the 50-V scale. (It is for this reason that you should select a range that yields a reading close to full scale when using a VOM.) In addition, VOM accuracies are usually specified with the meter lying in a horizontal position. Note also that accuracy specifications for ac measurements are generally poorer than for dc measurements.

FOR FURTHER INVESTIGATION

Analog meter readings are subject to parallax error. (This is a user reading error, not a meter error.) Go to the Internet and research parallax errors in meter reading. Prepare a short report for your instructor. Use diagrams to illustrate the parallax effect.

Safety

Everyone has handled electrical appliances and tools in one form or another, and most have never experienced a serious injury or mishap involving electricity. As a result, most of us think of electricity as a safe form of energy—if we think of it at all! However, we have all heard of someone getting a nasty shock as the result of carelessness or from poorly maintained equipment.

Although it is impossible to foresee every possible injury, most electrical accidents are preventable by using common sense and by adopting a healthy respect for electrical energy.

CURRENT KILLS!

The human body uses electrical impulses as low as 10 mV (ten one-thousandths of a volt) to transmit nerve messages between the brain and the various parts of the body. When an external source of electricity interacts with the body, the results can be disastrous! The human body consists mostly of water, with numerous dissolved electrolytes such as potassium chloride, phosphates, and sodium chloride, making us very good conductors of electricity. An alternating current as low as 5 mA (five one-thousandths of an ampere) can be quite painful, causing muscles to go into spasms. When the current is increased to 10 mA, the spasms may be sufficient to prevent the victim from letting go of the current source. If the current is increased to 15 mA, breathing may be stopped and the heart itself may go into spasms (fibrillation), preventing the flow of blood to the brain. If the victim does not receive immediate attention, death may result.

In the event of an electrical accident in the lab, there are some basic steps that you should take.

1. **Turn the power off if you can safely do so, otherwise (if you can safely do so), remove the victim from contact with the source of electricity.** Since touching the victim also puts the rescuer at risk, it is necessary to do this without coming into electrical contact. One way to do this is to use an insulated object such as a dry broom or other long, dry stick.
2. **Call for professional help.**
3. **If you are qualified, you can administer first aid.** You will need to tell the attending paramedics how the victim arrived at his/her injuries and what first aid was administered.

PREVENTING ELECTRICAL INJURY

Rather than treating an electrical injury, it is best to prevent the occurrence in the first place. The following is a list of safety rules which should be observed in the workplace or the lab.

1. **Consider all voltages above 50 Vdc or 30 Vac (rms) as a potential hazard.** If a voltage source is touched with dry hands you may experience only a mild tingle. The same source may be painful or fatal if your skin is moist or has a cut. If you experience even a minor electrical shock, mention this to your instructor or lab supervisor so that no one else is subjected to a potential hazard. It may be necessary to service the equipment.
2. **Ensure that all equipment is operating correctly and that no power cords are frayed or cut.** If a power cord is not in perfect condition, label the fault and bring it to the attention of your instructor or supervisor. Do not put a faulty piece of equipment into service. Tag the equipment and submit it for repair.
3. **Carefully insert and remove power cords.** Remove power cords by pulling on the connector. Never remove a power cord from the socket by pulling on the cable. There is a good chance that the cable will eventually tear free of the connector, presenting a fire or electrical hazard.
4. **Consider all sources of RF (radio frequency) energy as potential hazards.** Signals at very high frequencies (such as microwaves) have the potential to damage molecular structures (including your body's skin and internal organs) without immediately apparent symptoms.
5. **Use electrolytic capacitors with caution.** Some electrolytic capacitors are able to store large amounts of charge and may have the capacity to inflict the same injuries as voltage sources. Ensure that electrolytic capacitors are connected into the circuit with the correct polarity, and verify that the voltage rating of the capacitor is not exceeded. If an electrolytic capacitor is connected incorrectly, the result may be an explosion, sending electrolytic chemicals and sharp pieces of metal into the air. Safety glasses provide some protection from shrapnel and chemical spills.
6. **Use the correct protective equipment.** Prior to working on circuits with dangerous voltages, interrupt the circuit at the control panel by opening the switch. To prevent someone from inadvertently closing the switch, tag the circuit at the control panel, clearly stating the reason for the interrupt. Wear the appropriate protective gear such as safety glasses, gloves, and boots. Avoid wearing jewelry and loose-fitting clothing. Use a rubber mat and rubber-soled boots to work on circuits in a damp environment.
7. **Know the location of fire extinguishers, circuit breakers, and panic safety switches.** In the event of an accident, the power should be immediately turned off. If a fire has broken out, you may need to use a fire extinguisher. Ensure that the extinguisher is appropriate for the particular application. A fire should only be fought if it is not between you and a safe exit and if there is a good chance that you can bring the fire under control.

8. **Follow your instructor's safety instructions.** Since each lab has its own procedures which need to be followed, your instructor (or lab supervisor) will provide you with special precautions for the equipment or components used.

A NOTE CONCERNING CALCULATED AND MEASURED RESULTS

In the real world, calculated results and measured results seldom agree exactly. It is important to understand that there are usually explainable reasons for the differences, such as component tolerances, inherent meter errors, human measurement errors, and so on. As you analyze the data for each experiment, you should try to determine the explainable differences, and discuss them as part of the analysis for the lab.

TRADEMARKS

Multisim is a registered trademark of National Instruments Corporation.
OrCAD PSpice is a registered trademark of Cadence Design Systems, Inc.
Powerstat® is a registered trademark of The Superior Electric Co.
Excel is a registered trademark of Microsoft Corporation.

SUMMARY OF SOME SAFETY PRECAUTIONS WHEN USING LAB EQUIPMENT

Here are some general safety precautions to be observed for *all* lab work. You will also find specific safety advice in later labs. At all times, be careful.

1. Turn power off before making changes to your circuit.
2. Insert the probes into the meter before connecting the probes to your circuit in order to avoid "live" dangling leads.
3. Select the function and range (if necessary) of your meters before connecting to your circuit. If you do not have autoranging and do not know the approximate voltage or current that you are about to measure, set the meter to its highest range and work your way down to the appropriate range.
4. Keep your fingers behind the finger guards on the probes when making measurements.

Name ______________________________

Date ______________________________

Class ______________________________

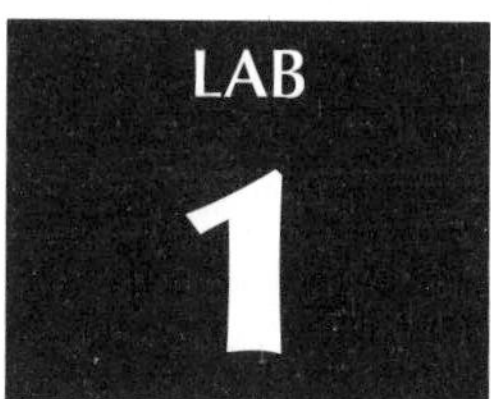

Voltage and Current Measurement and Ohm's Law

OBJECTIVES

After completing this lab, you will be able to
- measure voltage, current and resistance in a dc circuit,
- confirm Ohm's law by direct measurement.

EQUIPMENT REQUIRED

☐ Two digital multimeters (DMMs)
☐ Variable dc power supply

COMPONENTS

☐ Resistors: One 1-kΩ resistor, 1/4-W, 5% tolerance or better
Two 2-kΩ resistors, 1/4-W, 5% tolerance or better

PRELIMINARY

Before you start, review the section on measuring voltage, current, and resistance in *A Guide to Lab Equipment and Laboratory Measurements* at the front of this manual.

PRE-STUDY (OPTIONAL)

Interactive simulations *CircuitSim 02-2, 02-3,* and *04-1* from the Student Premium Website may be used as pre-study for this lab. If you have not already done so, go to the Student Premium Website, select *Learning Circuit Analysis Interactively on Your Computer,* download and unzip *Ch 2,* and run simulations *Sim 02-2* and *Sim 02-3*. Similarly download *Ch 4* and run *Sim 04-1*. (Be sure to select Multisim 11 or Multisim 9 files as applicable.) Instructions are provided.

EQUIPMENT USED

TABLE 1-1

Instrument	Manufacturer/Model No.	Serial No.
DMM #1		
DMM #2		
Power supply		

TEXT REFERENCE

Section 4.1 OHM'S LAW
Section 4.7 NONLINEAR AND DYNAMIC RESISTANCES

DISCUSSION

The most fundamental relationship of circuit theory is Ohm's law. Ohm's law states that in a purely resistive circuit, current is directly proportional to voltage and inversely proportional to resistance. In equation form

$$I = \frac{V}{R} \text{ (Ohm's law)} \qquad (1\text{-}1)$$

where V is in volts, R is in ohms, I is in amps, and reference conventions for voltage and current are as shown in Figure 1-1. This relationship (which holds for every resistance in a circuit) is an experimental result, and we will thus investigate it experimentally. First, you need to learn how to measure voltage and current.

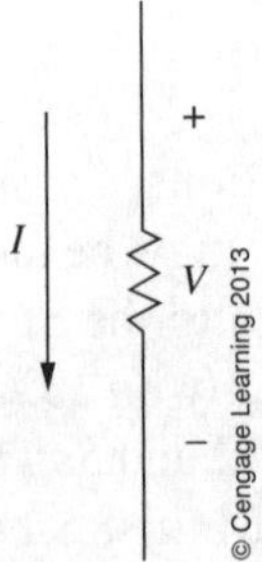

FIGURE 1-1 Conventions for Ohm's law.

Safety Checklist

Be sure to observe all safety precautions listed in *A Guide to Lab Equipment and Laboratory Measurements* in the front part of this manual.

dc Voltage Measurement

The basic voltage measuring scheme is shown in Figure 1-2. Be sure to adhere to the lead-color convention depicted and the safety checklist shown.

Current Measurement

The basic test circuit is depicted in Figure 1-3. When connected as shown, the DMM will yield a positive reading, whereas, if the test leads are reversed (so that current enters the COM terminal), the DMM will yield a negative reading.

MEASUREMENTS

PART A: Fixed Resistance, Variable Voltage

TABLE 1-2

	Nominal	Measured
R_1	1-kΩ	
R_2	2-kΩ	
R_3	2-kΩ	

1. Set your DMM to the ohms position, then measure the value of each resistor and record it in Table 1-2. (Mark each 2-kΩ resistor with masking tape so that you can keep track of them.)
2. With the power off, select R_2 and assemble the circuit as in Figure 1-4. Be sure to connect the meters so that they indicate positive values.

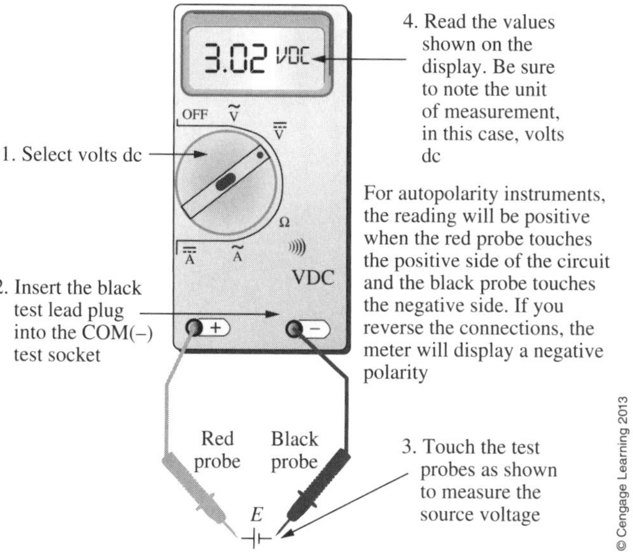

Safety Checklist

- Always turn power off before making changes to your circuit.
- Select the function (and range if necessary) before connecting the meter. If you do not have autoranging and you do not know the approximate voltage, begin with the highest range and work your way down.
- Keep your fingers behind the finger guards on the probes when making measurements. (Although this is not essential here because of the low voltages used, it is a good idea to develop safe work habits from the outset.)

FIGURE 1-2 Making a dc voltage measurement with a DMM. Steps 1 and 2 may be reversed.

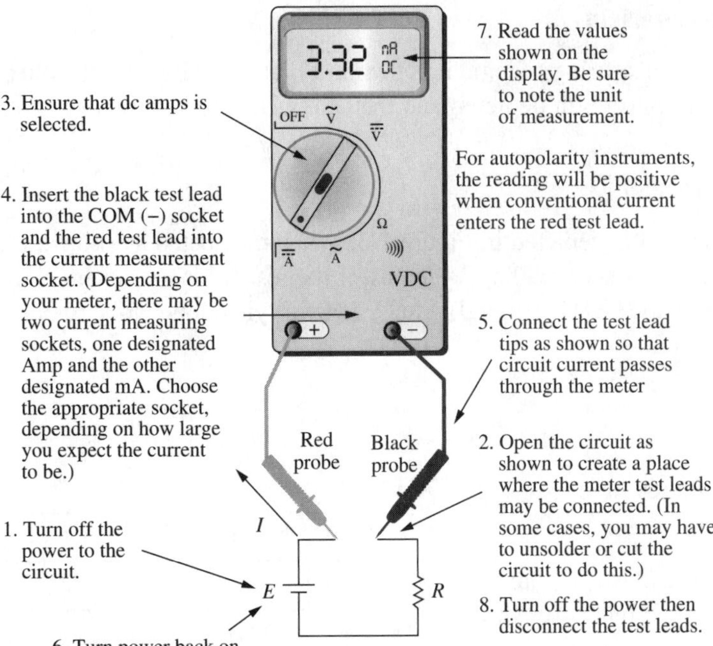

(a) Pictorial. Although a hand-held DMM is illustrated a bench-top DMM or a VOM may be used instead.

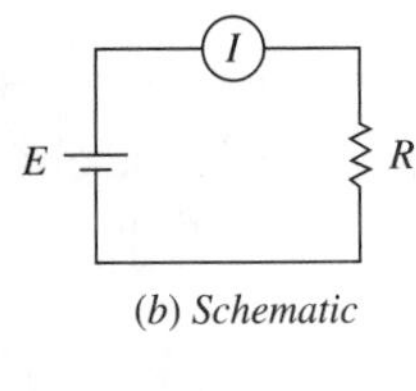

(b) Schematic

FIGURE 1-3 dc Current measurement.

Meter Terminal Designations

While the DMM shown has separate terminals for current and voltage, not all meters follow this practice. Some meters, for example, use a combined input designated VΩA, which means that the current input terminal is the same as the voltage input terminal. In this case, when you set the selector switch to A (or amps), the meter functions as an ammeter, whereas when you set it to V, it functions as a voltmeter. As always, connect the black lead to the COM terminal and the red lead to the other terminal.

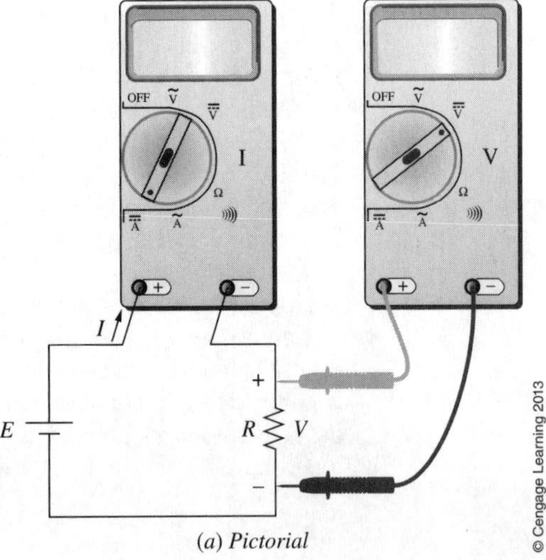
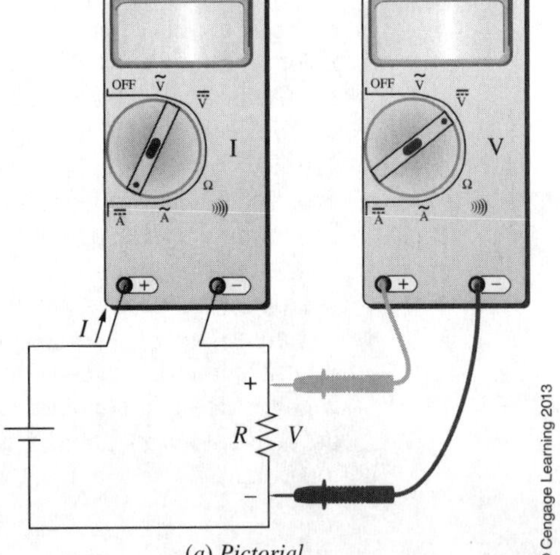

(a) Pictorial

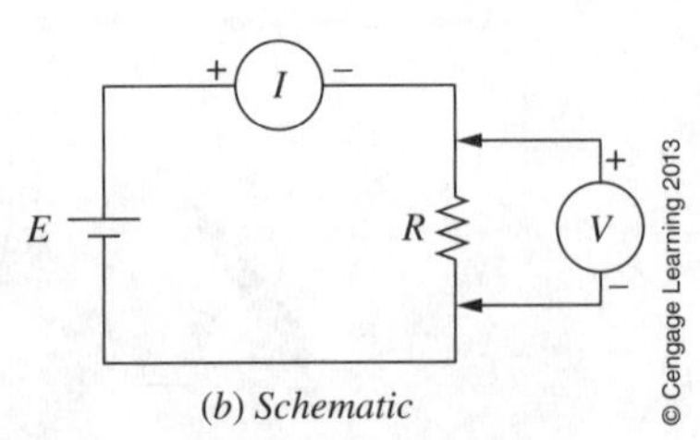

(b) Schematic

FIGURE 1-4 Connections for studying Ohm's law.

a. Set the voltage to $V = 8$ V and measure the current.

$I =$ _____________ (Record this result also in Table 1-3.)

b. Set the voltage to $V = 16$ V and measure the current.

$I =$ _____________

c. Set the voltage to $V = 4$ V and measure the current.

$I =$ _____________

d. Based on these observations, for a fixed resistance, how does current vary with voltage (within the limitations of accuracy of your meters and components)?

____

PART B: Fixed Voltage, Variable Resistance

3. a. Turn off the power supply. Using both 2-kΩ resistors, connect the circuit as in Figure 1-5. (This doubles the circuit resistance to 4 kΩ.) Set $V = 8$ V and measure the current.

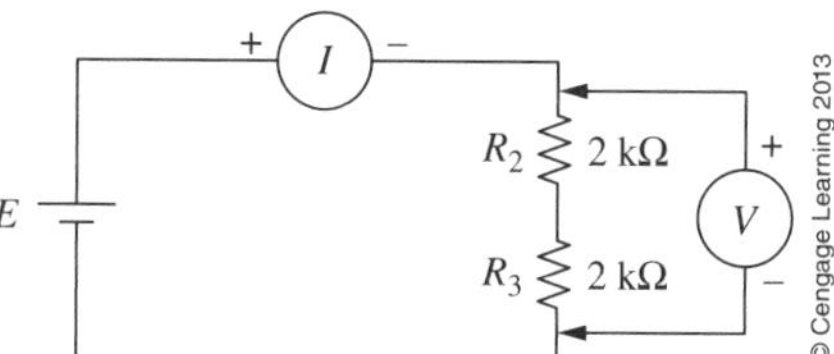

FIGURE 1-5 Doubling the circuit resistance. Here, $R_T = 4$ kΩ.

$I =$ _____________. (Record also in Table 1-3.)

b. Turn off the power supply. Using the 1-kΩ resistor, reconnect the circuit as in Figure 1-4. Set $V = 8$ V and measure the current.

$I =$ _____________. (Record also in Table 1-3.)

c. Consider Table 1-3. Within the accuracy of the results obtained, for a fixed voltage, how does current vary with resistance?

____

TABLE 1-3

	$V = 8$ V	
Test	R	I(mA)
2(a)	2 kΩ	
3(a)	4 kΩ	
3(b)	1 kΩ	

4. For Tests 2 and 3, compute current using Ohm's law and tabulate in Table 1-4. (Use the measured values of resistance for each case.) If there are differences between calculated and measured current, what is the likely cause? ____

____

TABLE 1-4

			Current I (mA)	
Test	V	R	Calculated	Measured
2(a)	8 V			
2(b)	16 V			
2(c)	4 V			
3(a)	8 V			
3(b)	8 V			

PART C: Ohm's Law Graph

5. Using the circuit of Figure 1-4 and the 1-kΩ resistor, vary V from 0 V to 10 V in 2-volt increments. Tabulate your results in Table 1-5 and plot the results on the graph of Figure 1-6. Replace R with a 2-kΩ resistor and repeat.
6. Resistance can be calculated at any point on the Ohm's law graph. For the nominal 1-kΩ resistor, at $V = 5$ V, determine current from Figure 1-6 and compute R using Equation 1-1. How well does it agree with the value of R used in Test 5? Repeat for the nominal 2-kΩ resistor.

____

____

____

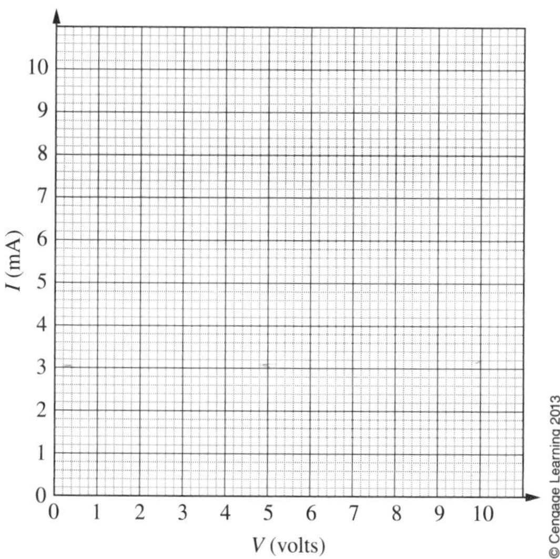

FIGURE 1-6 Ohm's law graph.

TABLE 1-5

	Current I (mA)	
V	1-kΩ	2-kΩ
0 V		
2 V		
4 V		
6 V		
8 V		
10 V		

7. Resistance may be determined from a *V-I* plot as described in Sec. 4.7 of the text. Select an arbitrary value for ΔV, then from the graph for the 1-kΩ resistor of Figure 1-6, determine the resulting ΔI. Using these values, compute resistance. Show all work, including a sketch with ΔV and ΔI clearly marked.

COMPUTER ANALYSIS

8. Using either Multisim or PSpice as applicable, set up the circuit of Figure 1-4. Repeat Part A and Part B of the lab and compare to your measured results.

PROBLEMS

9. A 12-V source is applied to a resistor with color code orange, black, brown. What is the current in mA?
10. Make a sketch to show how to connect voltmeters to measure the voltage across R_2 and R_3 of Figure 1-5.
11. For Figure 1-4, $I = 10$ mA. If E is increased by a factor of 16 and R is halved, what is the new value of current?

FOR FURTHER INVESTIGATION AND DISCUSSION

Write a short discussion paper on the impact of meter errors on measured values. As the basis for your discussion, consider a 4-digit DMM with an error specification of $\pm$(0.5% of reading + 1 digit) on its dc voltage ranges and $\pm$(1.5% of reading + 2 digits) on its dc current ranges. Consider Figure 1-4. Suppose the meters read 24.00 V and 12.00 mA, respectively. Compute the nominal value for resistance (assuming no meter errors) as well as the minimum and maximum resistance values assuming maximum meter errors. Based on this, what can you conclude about accepting results (derived from measurements) at face value?

Name ______________________

Date ______________________

Class ______________________

LAB 2

Series dc Circuits

OBJECTIVES

After completing this lab, you will be able to

- assemble a series circuit consisting of a voltage source and several resistors,
- use a DMM to measure voltage and current in a series circuit,
- compare measured values to theoretical calculations and verify Kirchhoff's voltage law,
- measure the effects of connecting several voltage sources in series,
- connect a circuit to ground using the ground terminal of a voltage source and use a DMM to measure voltages between several points and ground,
- measure the internal resistance of several voltage sources.

EQUIPMENT REQUIRED

☐ Digital multimeter (DMM)
☐ dc power supply (2)
Note: Record this equipment in Table 2-1.

COMPONENTS

☐ Resistors: 47-Ω, 100-Ω, 270-Ω, 330-Ω, 470-Ω (1/4-W)
47-Ω, 100-Ω, 270-Ω (2-W)
☐ Batteries: 1.5-V D-cell, 9-V (MN 1604 or equivalent)

EQUIPMENT USED

TABLE 2-1

Instrument	Manufacturer/Model No.	Serial No.
DMM		
dc Supply		
dc Supply		

TEXT REFERENCE

Section 5.1 SERIES CIRCUITS
Section 5.2 KIRCHHOFF'S VOLTAGE LAW
Section 5.3 RESISTORS IN SERIES
Section 5.4 VOLTAGE SOURCES IN SERIES
Section 5.5 INTERCHANGING SERIES COMPONENTS
Section 5.6 VOLTAGE DIVIDER RULE
Section 5.7 CIRCUIT GROUND
Section 5.9 INTERNAL RESISTANCE OF VOLTAGE SOURCES

DISCUSSION

Two elements are said to be in *series* if they are connected at a single point and if there are no other current-carrying connections at this point. Each element in a series circuit has the same current, as illustrated in Figure 2-1.

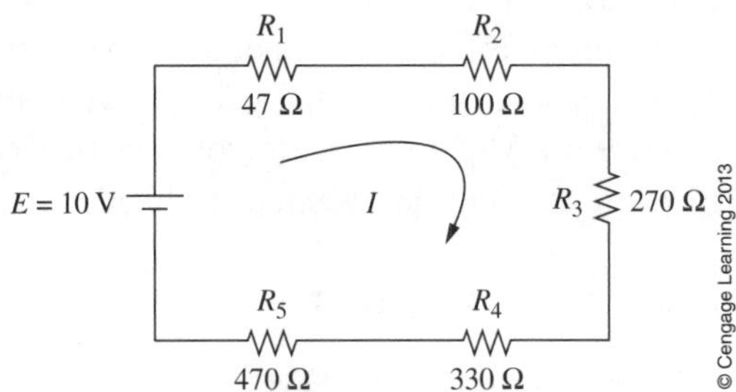

FIGURE 2-1 Series circuit.

The equivalent resistance of n resistors in series is determined as the summation

$$R_T = R_1 + R_2 + \ldots + R_n \qquad (2\text{-}1)$$

When these resistors are connected in series with a voltage source, the current in the circuit is given as

$$I = \frac{E}{R_T} \qquad (2\text{-}2)$$

The voltage drop across any resistor in a series circuit is determined by using the voltage divider rule, namely

$$V_x = \frac{R_x}{R_T} E \qquad (2\text{-}3)$$

CALCULATIONS

1. Refer to the circuit of Figure 2-1. Calculate R_T, I, and the voltage across each resistor. Enter the results in Table 2-2. Show the correct units for each entry.

TABLE 2-2

Resistor	Voltage
$R_1 = 47\ \Omega$	
$R_2 = 100\ \Omega$	
$R_3 = 270\ \Omega$	
$R_4 = 330\ \Omega$	
$R_5 = 470\ \Omega$	
R_T	
I	

MEASUREMENTS

2. Connect the resistors as shown in the network of Figure 2-2. Use the DMM (ohmmeter) to measure the resistance across the open terminals. Enter the result below. Compare your measurement to the theoretical calculation recorded in Table 2-2. You should observe only a small discrepancy (no greater than the percent tolerance of the resistors).

R_T	

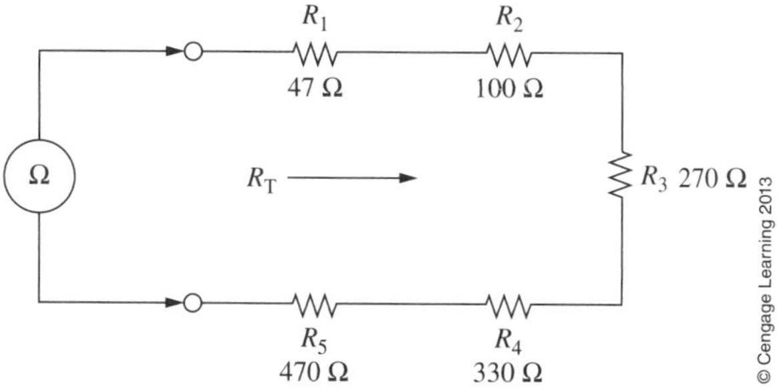

FIGURE 2-2 Series resistance.

3. Connect the voltage source into the circuit as shown in Figure 2-3. With a DMM (voltmeter) connected across the voltage source, adjust the voltage for exactly 10 V.

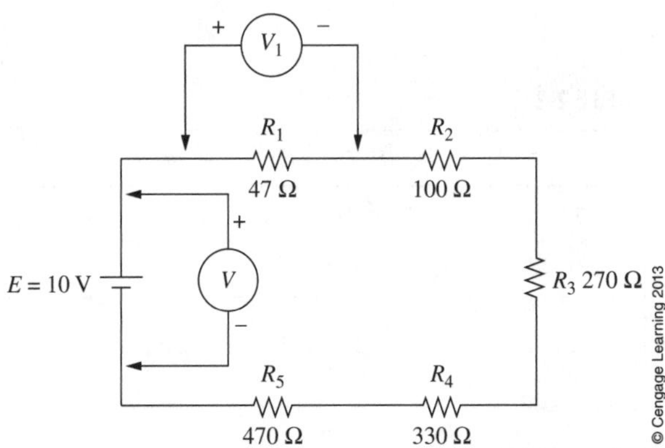

FIGURE 2-3 Measuring voltage in a series circuit.

4. Disconnect the voltmeter from the voltage source and successively place it across each of the resistors. Measure the voltage across each resistor in the circuit and record your results in Table 2-3.

TABLE 2-3

	Voltage
V_1	
V_2	
V_3	
V_4	
V_5	

5. Turn off the voltage source and switch the DMM from the voltage range to the current range. Disconnect the circuit and insert the ammeter in series with the circuit as shown in Figure 2-4.

Points to Note

The ammeter must be placed in series with the circuit, allowing the circuit current to pass through the meter. It must never be connected across an element, since this will result in a *short circuit* and may damage the meter or the circuit.

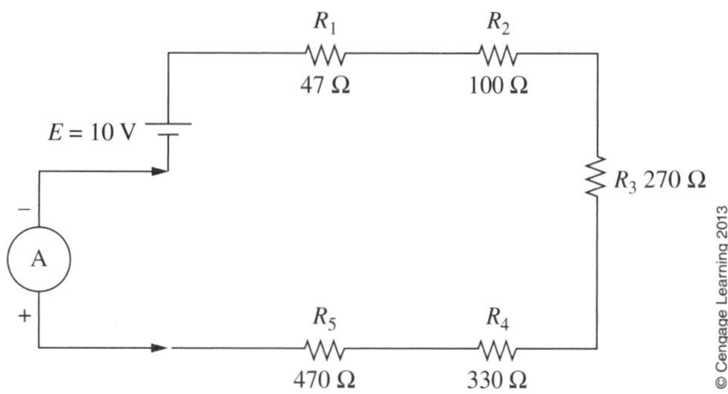

FIGURE 2-4 Current measurement.

Turn the voltage source back on and record the circuit current in the space provided here. If all resistors have a 5% tolerance, you should measure a current that is within 5% of the calculated value of Table 2-2.

I	

6. Obtain a second voltage source or use the second supply of a dual power supply. Modify the series circuit by placing the second source in a *series-aiding connection* with the first voltage source as illustrated in Figure 2-5. Adjust the second supply for 5 V. (The first supply is still 10 V.) Notice that in the series-aiding connection, the positive terminal of the second source is connected to the negative terminal of the first source. Measure and record in Table 2-4 the voltage drop across each resistor.

TABLE 2-4

	Voltage
V_1	
V_2	
V_3	
V_4	
V_5	

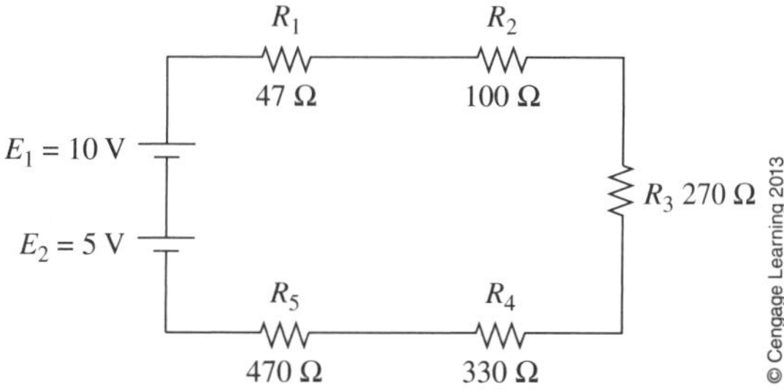

FIGURE 2-5 Voltage sources in a series-aiding connection.

7. Measure the current in the circuit using the method described in Step 5. Enter the result here.

I	

8. Turn both voltage sources off and reverse the terminals of the 5-V supply as shown in Figure 2-6. Turn both supplies on and adjust the voltages if necessary. The voltage sources are now in a *series-opposing connection*. Examine the circuit of Figure 2-6 and determine the correct direction of current. Measure and record in Table 2-5 the voltage drop across each resistor.

TABLE 2-5

	Voltage
V_1	
V_2	
V_3	
V_4	
V_5	

9. Measure the current in the circuit using the method described in Step 5. Enter the result here.

I	

Ground

Electrical and electronic circuits are often connected to *ground*, meaning that this part of the circuit is at the same potential as the ground connection on a three-terminal plug. Since all grounds are connected in a building's electrical panel (as well as the water pipes), special precautions may need to be followed

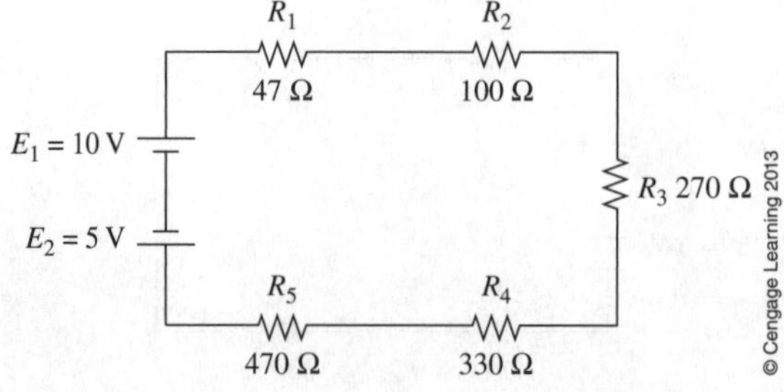

FIGURE 2-6 Voltage sources in a series-opposing connection.

when using instruments that are also grounded. Most dc voltage sources have a separate ground terminal at the front of the instrument, which permit us to connect a circuit to ground.

10. Refer to Figure 2-7(a). Connect the common terminals between the voltage sources to the ground terminal by using either a jumper wire or the grounding strip provided with the voltage source(s). Figure 2-7(b) indicates an alternate way of representing the voltage sources as *point sources*. Point sources simply indicate the potential at a point with respect to the reference (or in this case, ground). The ground symbol may or may not be shown. Connect the common (−) terminal of the DMM (voltmeter) to the ground point. All further voltage measurements are taken with respect to this point. Connect the voltage (+) terminal of the voltmeter to measure E_1, E_2, V_a, V_b, V_c, and V_d. Record your results in Table 2-6.

TABLE 2-6

	Voltage
E_1	
E_2	
V_a	
V_b	
V_c	
V_d	

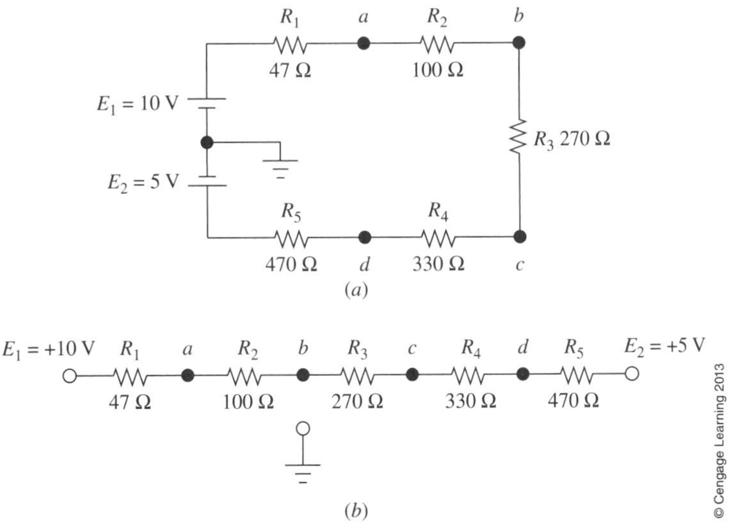

FIGURE 2-7 Ground connections and point sources.

Internal Resistance of Voltage Sources

All voltage sources have some internal resistance, which tends to reduce the voltage between the terminals of the source when the source is under load. The internal resistance of the voltage sources used up to now has been very small (typically less than 1 Ω). Other sources such as nickel-cadmium and alkaline batteries have relatively large internal resistance. This means that as a circuit requires more current, the voltage across the terminals of the battery will decrease. The magnitude of the internal resistance determines the maximum amount of usable current that a voltage source can provide to a circuit.

11. Use a DMM to measure the voltage across the terminals of a 1.5-V D-cell alkaline battery and a 9-V alkaline battery. Enter the results here.

E(1.5-V cell)	
E(9-V cell)	

12. Using each of the resistors from the previous part of the lab, construct the circuit shown in Figure 2-8. Measure and record the voltage across the terminals of the batteries for each of the resistors in Table 2-7.

Caution

When connecting some of the resistors across the 9-V battery, the 1/4-W power rating will be exceeded. You will need to use 2-W resistors to permit voltage measurements with the 47-Ω, 100-Ω, and 270-Ω resistors. Since the resistors may get quite hot, extra precaution should be observed when handling them.

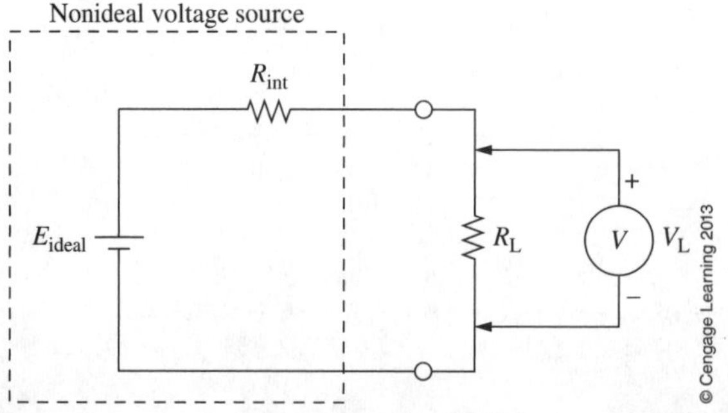

FIGURE 2-8 Measuring internal resistance of a voltage source.

TABLE 2-7

	V_L (1.5-V cell)	V_L (9-V cell)
$R_L = 470\ \Omega$		
$R_L = 330\ \Omega$		
$R_L = 270\ \Omega$		
$R_L = 100\ \Omega$		
$R_L = 47\ \Omega$		

CONCLUSIONS

13. Compare the measured resistance of Step 2 to the theoretical resistance recorded in Table 2-2. Determine the percentage variation as shown.

R_T (theoretical) = ______________ R_T (measured) = ______________

$$\text{percent variation} = \frac{\text{Measurement} - \text{Theoretical}}{\text{Theoretical}} \times 100\% \qquad (2\text{-}4)$$

percent variation = ______________

14. Compare the measured voltage drops of Table 2-3 to the theoretical values recorded in Table 2-2. Indicate which values (if any) have a variation more than the resistor tolerance. Offer an explanation.

____

____

15. Examine the measured current of Step 5 and compare it to the theoretical value recorded in Table 2-2. Determine the percentage variation.

percent variation = ______________

16. Determine the summation of the voltage drops recorded in Table 2-4. Compare this value to the summation of voltage rises. Is Kirchhoff's voltage law satisfied?

ΣV = ______________ ΣE = ______________

17. Calculate the theoretical current for the circuit of Figure 2-5. Compare this value to the measured value and determine the percent variation.

$I_{\text{theoretical}} =$ ____________ percent variation = ____________

18. Determine the summation of the voltage drops recorded in Table 2-5. Compare this value to the summation of voltage rises. Is Kirchhoff's voltage law satisfied?

$\Sigma V =$ ____________ $\Sigma E =$ ____________

____

____

19. Calculate the theoretical current for the circuit of Figure 2-6. Compare this value to the measured value and determine the percent variation.

$I_{\text{theoretical}} =$ ____________ percent variation = ____________

20. Calculate the voltages V_{ab}, V_{bc}, and V_{cd} from your measurements in Table 2-6. Compare your calculations to the corresponding measurements recorded in Table 2-5. Notice the direct correlation.

$V_{ab} = V_a - V_b =$ ____________ $V_2 =$ ____________

$V_{bc} =$ ____________ $V_3 =$ ____________

$V_{cd} =$ ____________ $V_4 =$ ____________

21. For each load resistor, evaluate the internal resistance of the 1.5-V battery by applying Ohm's law as shown below. Enter your results in Table 2-8. Use these results to determine the average value of internal resistance.

$$I = \frac{V_L}{R_L} \tag{2-5}$$

$$R_{\text{int}} = \frac{E_{\text{ideal}} - V_L}{I} = \left(\frac{E_{\text{ideal}} - V_L}{V_L}\right)R_L \tag{2-6}$$

TABLE 2-8

R_L	R_{int}
470 Ω	
330 Ω	
270 Ω	
100 Ω	
47 Ω	
Average value of R_{int}	

22. For each load resistor, use Equations 2-5 and 2-6 to evaluate the internal resistance of the 9-V battery. Enter your results in Table 2-9. Use these results to determine the average value of internal resistance.

TABLE 2-9

R_L	R_{int}
470 Ω	
330 Ω	
270 Ω	
100 Ω	
47 Ω	
Average value of R_{int}	

FOR FURTHER INVESTIGATION AND DISCUSSION

Use Multisim or PSpice to simulate the circuit of Figure 2-3. Determine the voltage drop across each resistor in the circuit and compare your results to those of Table 2-3. (Note: In order simulate the circuit, you will need to select a suitable reference point.)

Name ______________________

Date ______________________

Class ______________________

LAB 3

Parallel dc Circuits

OBJECTIVES

After completing this lab, you will be able to

- assemble a parallel circuit consisting of a voltage source and several resistors,
- measure voltage and current in a parallel circuit,
- compare measured values to theoretical calculations and verify Kirchhoff's current law.

EQUIPMENT REQUIRED

☐ Digital multimeter (DMM)
☐ dc power supply
Note: Record this equipment in Table 3-1.

COMPONENTS

☐ Resistors: 470-Ω, 680-Ω, 1-kΩ, 2.2-kΩ, 4.7-kΩ (1/4-W, 5%)

EQUIPMENT USED

TABLE 3-1

Instrument	Manufacturer/Model No.	Serial No.
DMM		
dc Supply		

TEXT REFERENCE

Section 6.1 PARALLEL CIRCUITS
Section 6.2 KIRCHHOFF'S CURRENT LAW
Section 6.3 RESISTORS IN PARALLEL
Section 6.5 CURRENT DIVIDER RULE
Section 6.6 ANALYSIS OF PARALLEL CIRCUITS

DISCUSSION

Two elements are said to be in a *parallel* connection if they have exactly two nodes in common. Each element in a parallel circuit, as shown in Figure 3-1, has the same voltage across it.

The equivalent conductance of n resistors in parallel is determined as the summation of conductance

$$G_T = G_1 + G_2 + \dots + G_n \tag{3-1}$$

where the conductance G of each resistor is found as the reciprocal of resistance

$$G_x = \frac{1}{R_x} \tag{3-2}$$

The total resistance of n resistors in parallel is then found as

$$R_T = \frac{1}{G_T} \tag{3-3}$$

When a parallel network of resistors is connected in parallel with a voltage source, the current through the voltage source is determined as

$$I = \frac{E}{R_T} \tag{3-4}$$

The current through any resistor in a parallel circuit is calculated using Ohm's law or the current divider rule, namely

$$I_x = \frac{E}{R_x} = \frac{R_T}{R_x} I \tag{3-5}$$

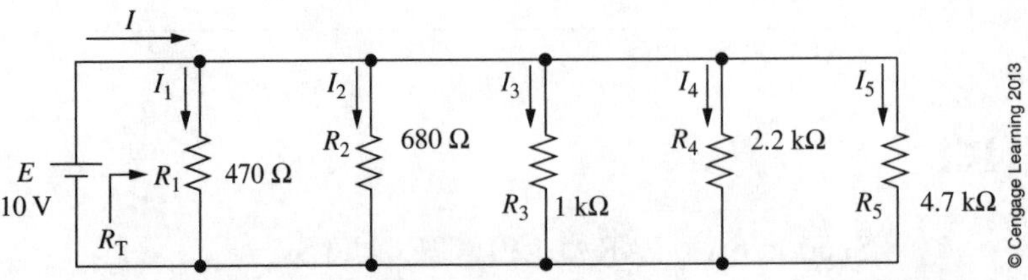

FIGURE 3-1 Parallel circuit.

CALCULATIONS

1. Refer to the circuit of Figure 3-1. Calculate R_T, I, I_1, I_2, I_3, I_4, and I_5. Enter the results in Table 3-2. Show the correct units for each entry.

TABLE 3-2

I_1	
I_2	
I_3	
I_4	
I_5	

R_T	
I	

© Cengage Learning 2013

MEASUREMENTS

2. Connect the resistors as shown in the network of Figure 3-2. Use the DMM (ohmmeter) to measure the resistance across the open terminals. Enter the result here. Compare your measurement to the theoretical calculation in Table 3-2. You should observe only a small discrepancy.

R_T	

3. Connect the voltage source to the circuit as shown in Figure 3-3. With a DMM (voltmeter) connected across the voltage source, adjust the voltage for exactly 10 V.

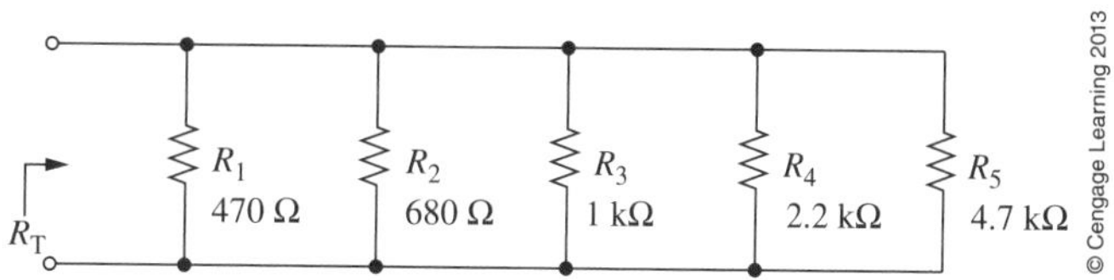

FIGURE 3-2 Parallel resistance.

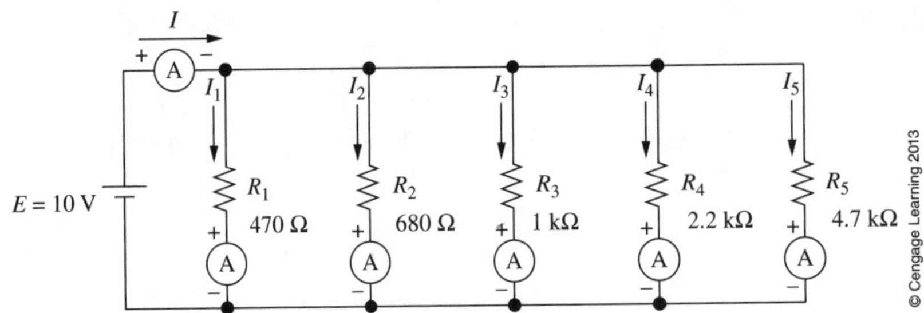

FIGURE 3-3 Measuring current in a parallel circuit.

4. Disconnect the voltmeter from the voltage source. Turn the power supply off, without disturbing the voltage setting on the power supply. Set the DMM to measure current. Successively disconnect each branch of the circuit and insert the ammeter into the branch as illustrated in Figure 3-3. Measure the current through the voltage source and through each resistor in the circuit. **Ensure that each branch is reconnected after the ammeter is removed.** Record your results in Table 3-3.

TABLE 3-3

	Current
I	
I_1	
I_2	
I_3	
I_4	
I_5	

© Cengage Learning 2013

KIRCHHOFF'S CURRENT LAW

Kirchhoff's voltage and current laws provide an important foundation for the analysis of circuits. Kirchhoff's current law states:

The summation of currents entering a node is equal to the summation of currents leaving the node.

We now examine how Kirchhoff's current law can be verified in a laboratory.

5. Relocate the ammeter as shown in Figure 3-4 and measure the currents I_6, I_7, and I_8. Record your results in Table 3-4.

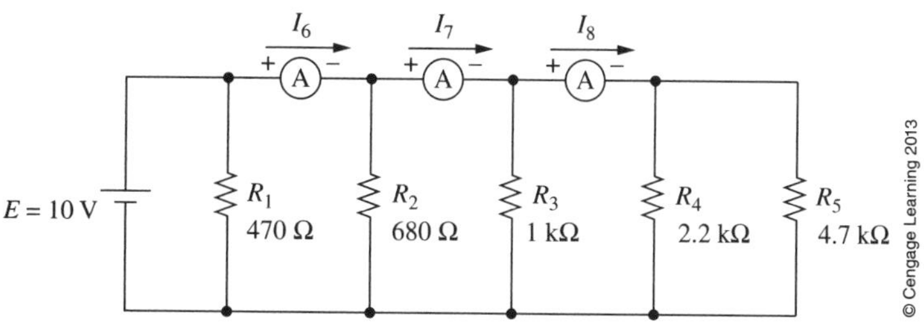

FIGURE 3-4 Verifying Kirchhoff's current law.

TABLE 3-4

	Current
I_6	
I_7	
I_8	

CONCLUSIONS

6. Compare the measured resistance of Step 2 to the theoretical resistance recorded in Table 3-2. Determine the percentage variation as shown.

R_T (theoretical) = ______________ R_T (measured) = ______________

$$\text{percent variation} = \frac{\text{Measurement} - \text{Theoretical}}{\text{Theoretical}} \times 100\% \qquad (3\text{-}6)$$

percent variation = ____________

7. Compare the measured currents of Step 4 to the theoretical values recorded in Table 3-2. Indicate which values (if any) have a variation more than the resistor tolerance. Offer an explanation.

____

____

8. Refer to the data of Table 3-3 and Table 3-4.
   a. Compare current I to the summation $I_1 + I_2 + I_3 + I_4 + I_5$.

      $I =$ ________________ $I_1 + I_2 + I_3 + I_4 + I_5 =$ ________________

   b. Compare current I_6 to the summation $I_2 + I_3 + I_4 + I_5$.

      $I_6 =$ ________________ $I_2 + I_3 + I_4 + I_5 =$ ________________

   c. Compare current I_7 to the summation $I_3 + I_4 + I_5$.

      $I_7 =$ ________________ $I_3 + I_4 + I_5 =$ ________________

   d. Compare current I_8 to the summation $I_4 + I_5$.

      $I_8 =$ ________________ $I_4 + I_5 =$ ________________

   e. Do the above calculations and measurements verify Kirchhoff's current law? Explain your answer.

      ____

      ____

FOR FURTHER INVESTIGATION AND DISCUSSION

Use Multisim or PSpice to simulate the circuit of Figure 3-1. Determine the current through each resistor in the circuit. Compare your results to those of Table 3-2.

Name ______________________

Date ______________________

Class ______________________

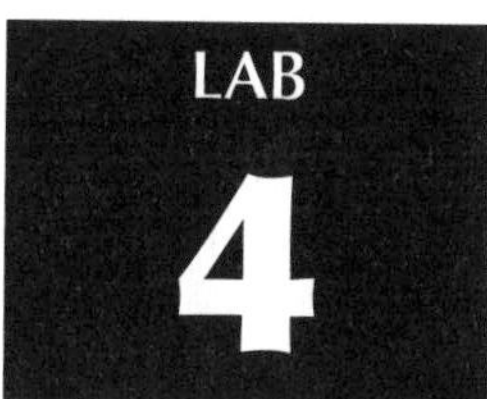

Series-Parallel dc Circuits

OBJECTIVES

After completing this lab, you will be able to

- assemble a series-parallel circuit consisting of a voltage source and several resistors,
- measure voltage and current in a series-parallel circuit,
- compare measured values to theoretical calculations and verify Kirchhoff's current and voltage laws,
- assemble a zener diode regulator circuit and measure voltages and currents to verify that Kirchhoff's voltage and current laws apply,
- calculate power and verify the law of conservation of energy.

EQUIPMENT REQUIRED

☐ Digital multimeter (DMM)
☐ dc power supply
Note: Record this equipment in Table 4-1.

COMPONENTS

☐ Resistors: 330-Ω, 470-Ω, 680-Ω, 1-kΩ, 2.2-kΩ, 4.7-kΩ (1/4-W, 5%)
☐ Zener diode: 1N4734A (1-W, 5.6-V5%)

EQUIPMENT USED

TABLE 4-1

Instrument	Manufacturer/Model No.	Serial No.
DMM		
dc Supply		

© Cengage Learning 2013

TEXT REFERENCE

Section 7.1 THE SERIES-PARALLEL NETWORK
Section 7.2 ANALYSIS OF SERIES-PARALLEL CIRCUITS
Section 7.3 APPLICATIONS OF SERIES-PARALLEL CIRCUITS

DISCUSSION

Regardless of the complexity of a circuit, the basic laws of circuit analysis always apply. While Ohm's law and Kirchhoff's voltage and current laws are used to analyze simple series and parallel circuits, these same laws may be applied to analyze even the most complicated circuit. The following rules apply to all circuits:

The same current occurs through all series elements.

The same voltage appears across all parallel elements.

CALCULATIONS

1. Refer to the circuit of Figure 4-1. Calculate the total resistance, R_T, seen by the voltage source. Calculate the current, I. Solve for all resistor currents, voltages, and powers. Enter the results in Table 4-2. Show the correct units for all entries.

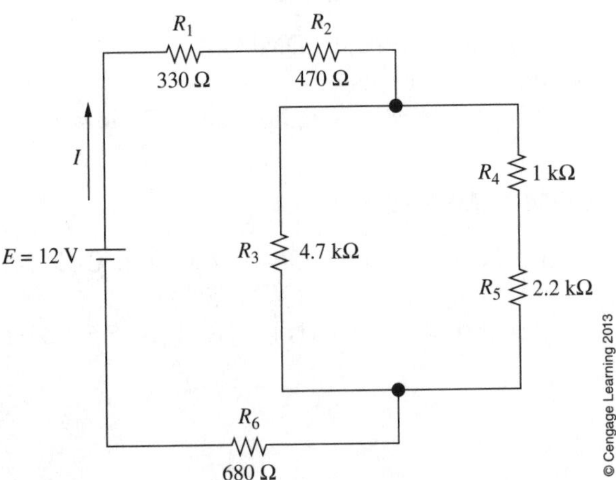

FIGURE 4-1 Series-parallel circuit.

TABLE 4-2 Series-Parallel Resistance

	Current	Voltage	Power
R_1			
R_2			
R_3			
R_4			
R_5			
R_6			

R_T	
I	

MEASUREMENTS

2. Connect the resistors as shown in the network of Figure 4-2. Use the DMM (ohmmeter) to measure the resistance across the open terminals. Enter the result here. Compare your measurement to the theoretical calculation in Table 4-2. You should observe only a small discrepancy.

R_T	

3. Connect the voltage source to the circuit as shown in Figure 4-1. With a DMM (voltmeter) connected across the voltage source, adjust the voltage for exactly 12 V.

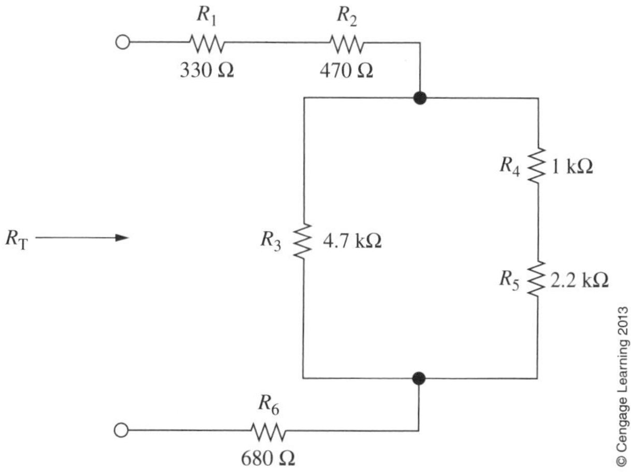

FIGURE 4-2 Series-parallel resistance.

4. Disconnect the voltmeter from the voltage source. Measure the voltage across each resistor in the circuit and record the results in Table 4-3.
5. Set the DMM to measure the currents as illustrated in Figure 4-3. **Ensure that each branch is reconnected after the ammeter is removed.** Record your measurements in Table 4-3.

TABLE 4-3

	Current	Voltage
R_1		
R_2		
R_3		
R_4		
R_5		
R_6		

I	

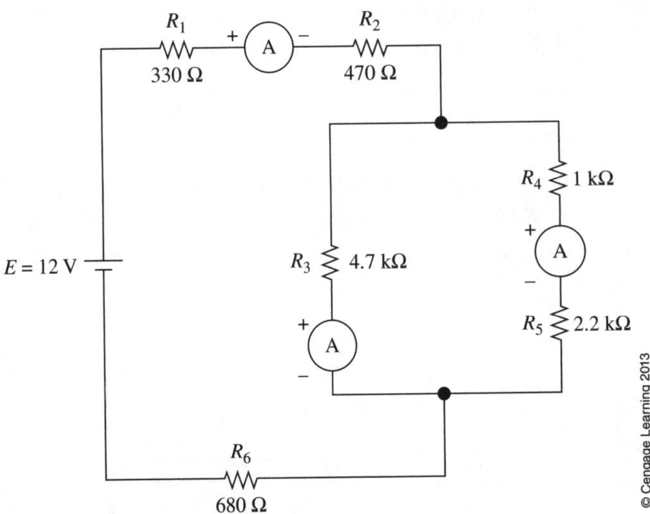

FIGURE 4-3 Measuring current in a series-parallel circuit.

Zener Diode Circuit

In this part of the lab, we apply the principles of circuit analysis to examine the operation of a more complicated circuit. Here we use a *zener diode,* which is a two-terminal semiconductor device normally used as a *voltage regulator* to maintain a constant voltage between two terminals. When the zener diode is placed across a component which has a voltage greater than the *break-over* (zener) voltage of the diode, current through the zener diode forces the voltage across the component to decrease. While most other diodes permit current in only the forward direction (in the direction of the arrow in the diode symbol), the zener diode can conduct in either direction. When used as a voltage regulator, the zener diode is operated in the reverse-biased condition. This means that when the zener diode is in its *break-over region*, conventional current is against the arrow of the diode symbol.

6. Assemble the circuit shown in Figure 4-4, temporarily omitting the zener diode. Adjust the voltage source for 12 V.
7. Measure the voltages across R_1 and R_2 and record the values in Table 4-4.

TABLE 4-4

V_1	
V_2	

8. Insert the zener diode into the circuit. Use the DMM (voltmeter) to measure voltages V_1 and V_2 in the circuit of Figure 4-4. Convert the DMM to measure current and correctly measure currents I_1, I_2, and I_Z. Make sure that you turn off the voltage supply before disconnecting the circuit to insert the ammeter. In the circuit of Figure 4-4, show where you placed the ammeters to measure the currents. Record all measurements in Table 4-5.

TABLE 4-5

Current	Voltage
I_1 =	V_1 =
I_2 =	V_2 =
I_Z =	

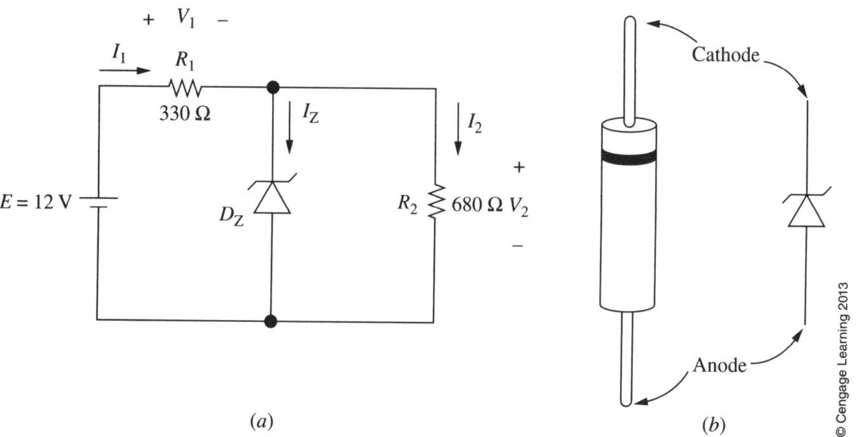

FIGURE 4-4 Zener diode voltage regulator circuit.

CONCLUSIONS

9. Compare the measured resistance of Step 2 to the theoretical value recorded in Table 4-2. Determine the percent variation.

R_T (theoretical) = ______________ R_T (measured) = ______________

percent variation = ______________

10. Compare the measured voltage drops in Table 4-3 to the theoretical values recorded in Table 4-2. Indicate which values (if any) have a variation more than the resistor tolerance. Offer an explanation.

____

____

11. Compare the measured currents in Table 4-3 to the theoretical values recorded in Table 4-2. Indicate which values (if any) have a variation more than the resistor tolerance. Offer an explanation.

____

____

12. Refer to the data of Table 4-3.

    a. Compare current I to the summation $I_3 + I_4$.

    I = ______________ $I_3 + I_4$ = ______________

    b. Do the above calculations and measurements satisfy Kirchhoff's current law? Explain your answer.

13. a. Use the data of Table 4-3 to calculate the power P_T delivered to the circuit by the voltage source. Enter the result here.

P_T	

    b. Use the voltages and currents of Table 4-3 to determine the power dissipated by each resistor in the circuit. Enter your results in Table 4-6.

TABLE 4-6

	Power
R_1	
R_2	
R_3	
R_4	
R_5	
R_6	

c. Compare the total power dissipated by the resistors to the total power delivered by the voltage source. Is energy conserved?

____

____

14. Refer to the circuit of Figure 4-4. Compare the voltage, V_2, with the zener diode removed from the circuit, to the voltage when the diode is in the circuit. How do they compare? Explain briefly why this occurred.

____

____

15. Refer to the data of Table 4-5.
 a. Compare current I_1 to the summation $I_2 + I_Z$.

$I_1 =$ ______________ $I_2 + I_Z =$ ______________

 b. Do the calculations and measurements in part (a) satisfy Kirchhoff's current law? Explain your answer.

____

____

16. a. Use the data in Table 4-5 to calculate the total power P_T delivered to the circuit by the voltage source. Enter the result here.

P_T	

b. Use the voltages and currents of Table 4-5 to determine the power dissipated by the zener diode and by each resistor in the circuit. Enter your results in Table 4-7.

TABLE 4-7

	Power
R_1	
R_2	
D_Z	

c. Compare the total power dissipated by the resistors to the total power delivered by the voltage source. Is energy conserved?

____

____

PROBLEMS

17. Refer to the circuit of Figure 4-4. Assume that the zener diode has a break-over voltage of 4.3 V.
    a. Calculate the voltages V_1 and V_2.
    b. Determine the currents I_1, I_2, and I_Z.
    c. Solve for the powers dissipated by R_1, R_2, and D_Z.
    d. Show that the total power dissipated is equal to the power delivered by the voltage source.

Name ______________________

Date ______________________

Class ______________________

LAB 5

Potentiometers and Rheostats

OBJECTIVES

After completing this lab, you will be able to

- demonstrate the use of a variable resistor as a potentiometer to control the voltage applied to a load,
- demonstrate the use of a variable resistor as a rheostat to control the current applied to a load,
- measure how the value of a load's resistance affects the voltage across a potentiometer.

EQUIPMENT REQUIRED

☐ Digital multimeter (DMM)
☐ dc power supply
Note: Record this equipment in Table 5-1.

COMPONENTS

☐ Resistors: 5.6-kΩ, 3.3-kΩ, 330-kΩ (1/4-W, 5%)
10-kΩ variable resistor

EQUIPMENT USED

TABLE 5-1

Instrument	Manufacturer/Model No.	Serial No.
DMM		
dc Supply		

TEXT REFERENCE

Section 3.5 TYPES OF RESISTORS
Section 7.4 POTENTIOMETERS

DISCUSSION

Variable resistors are used extensively in electrical and electronic circuits to control the voltage and current in circuits. When a variable resistor is used to control the voltage (as in the volume control of an amplifier) it is called a *potentiometer*. If the same resistor is used to control the amount of current through a circuit (such as in a light dimmer) it is called a *rheostat*.

Refer to the series circuit of Figure 5-1. The current through this circuit is constant regardless of the location of the wiper arm (terminal *b*) of the variable resistor. If the wiper arm is moved so that it is at the bottom of the resistor, the resistance between terminals *b* and *c* is zero. This results in the voltage V_L being zero volts. If, however, the wiper arm is moved so that it is at the top of the resistor, the resistance between terminals *b* and *c* will be at a maximum, resulting in a maximum voltage, V_L, appearing between the terminals.

CALCULATIONS

1. Determine the range of the output voltage V_L for the circuit of Figure 5-1. Calculate the range of output voltage if a 330-kΩ resistor is connected across the output terminals of the circuit. Recalculate the range of output voltage if a 3.3-kΩ resistor is connected across the output terminals of the circuit. Record your results in Table 5-2.

TABLE 5-2

	V_L (min)	V_L (max)
$R_L = \infty$ (open)		
R_L = 330 kΩ		
R_L = 3.3 kΩ		

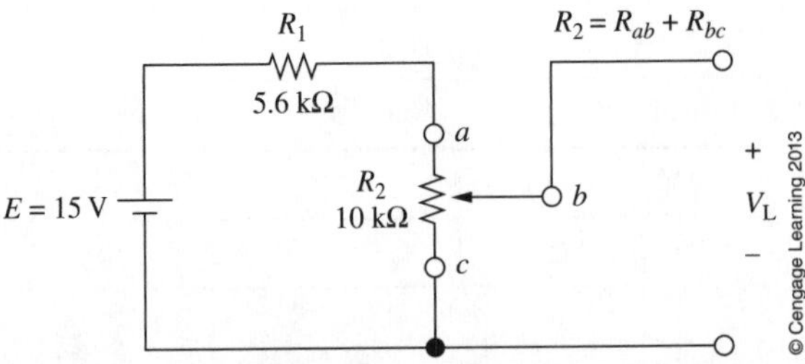

FIGURE 5-1 Variable resistor used as a potentiometer.

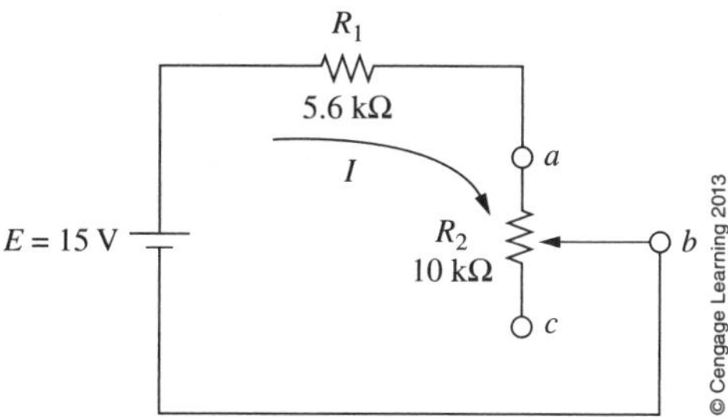

FIGURE 5-2 Variable resistor used as a rheostat.

2. The 10-kΩ resistor in the circuit of Figure 5-1 is easily converted from a potentiometer into a rheostat. Figure 5-2 shows how the variable resistor is used as a rheostat. When the wiper arm is moved so that it is at the top of the resistor, the resistance of the rheostat will be at its minimum value. This means that maximum current will occur in the circuit. If the wiper is moved to the bottom of the resistor, the resistance of the rheostat will be at its maximum value, resulting in the least amount of current.

   Calculate the range of current for the circuit of Figure 5-2.

I(min)	I(max)

MEASUREMENTS

3. Assemble the circuit shown in Figure 5-1. Adjust the voltage source for 15 V. With a voltmeter connected between terminals *b* and *c* of the potentiometer, use a small screwdriver to adjust the central wiper fully clockwise (CW). Measure and record the output voltage. Now adjust the central wiper fully counterclockwise (CCW). Again measure and record the output voltage.

V_L(CW)	V_L(CCW)

4. Adjust the potentiometer to obtain an output voltage of 3.0 V. Disconnect the potentiometer from the circuit, being careful not to readjust the potentiometer setting. Measure and record the resistance between terminals *b* and *c* of the potentiometer.

R_{bc}	

5. Connect a 330-kΩ load resistor between the output terminals. Adjust the voltage source for 15 V. With a voltmeter connected between terminals *b* and *c* of the potentiometer, use a small screwdriver to adjust the central wiper fully clockwise (CW). Measure and record the output voltage. Now adjust the central wiper fully counterclockwise (CCW). Again measure and record the output voltage.

V_L(CW)	V_L(CCW)

6. Adjust the potentiometer to obtain an output voltage of 3.0 V. Disconnect the potentiometer from the circuit, being careful not to readjust the potentiometer setting. Measure and record the resistance between terminals *b* and *c* of the potentiometer.

R_{bc}	

7. Connect a 3.3-kΩ load resistor between the output terminals. Adjust the voltage source for 15 V. With a voltmeter connected between terminals *b* and *c* of the potentiometer, use a small screwdriver to adjust the central wiper fully clockwise (CW). Measure and record the output voltage. Now adjust the central wiper fully counterclockwise (CCW). Again measure and record the output voltage.

V_L(CW)	V_L(CCW)

8. Adjust the potentiometer to obtain an output voltage of 3.0 V. Disconnect the potentiometer from the circuit, being careful not to readjust the potentiometer setting. Measure and record the resistance between terminals *b* and *c* of the potentiometer.

R_{bc}	

9. Construct the circuit of Figure 5-2. Place a DMM ammeter into the circuit to measure the current. Use a small screwdriver to adjust the central wiper fully clockwise (CW). Measure and record the circuit current. Now adjust the central wiper fully counterclockwise (CCW). Again measure and record the current.

I(CW)	I(CCW)

10. Adjust the rheostat so that the measured current is exactly equal to $I_{max}/2$. Remove the rheostat from the circuit, being careful not to readjust the potentiometer setting. Measure and record the resistance between terminals a and b of the potentiometer. The resistance should be exactly equal to R_1.

R_{ab}	

CONCLUSIONS

11. Compare the measurements of Step 3 to the theoretical maximum and minimum values of voltage recorded in Table 5-2 for $R_L = \infty$ (open).

12. Compare the measurements of Step 5 to the theoretical maximum and minimum values of voltage recorded in Table 5-2 for $R_L = 330\ k\Omega$.

13. Use the measured resistance R_{bc} of Step 6 to determine the theoretical voltage which would appear across the load $R_L = 330\ k\Omega$ in the equivalent circuit of Figure 5-3.

V_L	

Compare the above value to the measured load voltage V_L in Step 6.

14. Compare the measurements of Step 7 to the theoretical maximum and minimum values of voltage recorded in Table 5-2 for $R_L = 3.3\ k\Omega$.

15. Use the measured resistance R_{bc} of Step 8 to determine the theoretical voltage which would appear across the load $R_L = 3.3\ k\Omega$ in the equivalent circuit of Figure 5-3.

V_L	

Compare the above value to the measured load voltage V_L in Step 8.

16. Compare the measurements of Step 9 to the theoretical maximum and minimum values of current recorded in Step 2.

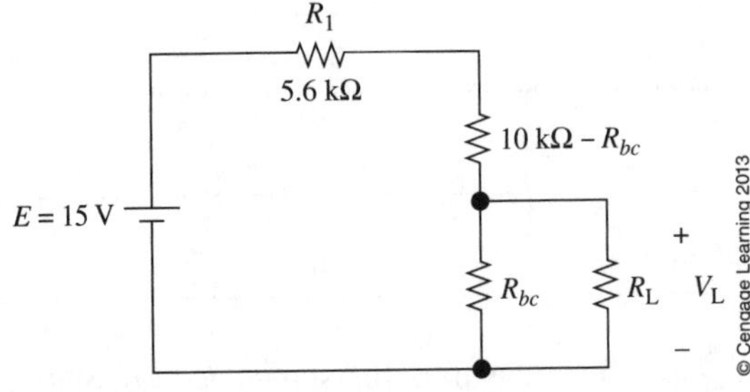

FIGURE 5-3

Name ______________________

Date ______________________

Class ______________________

Superposition Theorem

OBJECTIVES

After completing this lab, you will be able to

- calculate currents and voltages in a dc circuit using the superposition theorem,
- measure voltage and current in a multi-source circuit,
- measure the effects of successively removing each voltage source from a circuit,
- calculate loop currents and node voltages using mesh analysis and nodal analysis,
- verify the superposition theorem as it applies to dc circuits and show that the results are consistent with results determined using mesh analysis and nodal analysis.

EQUIPMENT REQUIRED

☐ Digital multimeter (DMM)
☐ dc power supply (2)
Note: Record this equipment in Table 6-1.

COMPONENTS

☐ Resistors: 680-Ω, 1-kΩ, 3.3-kΩ (1/4-W, 5%)

EQUIPMENT USED

TABLE 6-1

Instrument	Manufacturer/Model No.	Serial No.
DMM		
dc Supply		
dc Supply		

TEXT REFERENCE

Section 8.2 SOURCE CONVERSIONS
Section 8.5 MESH (LOOP) ANALYSIS
Section 8.6 NODAL ANALYSIS
Section 9.1 SUPERPOSITION THEOREM

DISCUSSION

Mesh analysis allows us to find the loop currents for a circuit having any number of voltage or current sources. If a circuit contains current sources, these must first be converted to voltage sources.

Nodal analysis is the twin of mesh analysis in that it allows us to calculate nodal voltages of a circuit (with respect to a reference node). If a circuit contains voltage sources, it is necessary to first convert these to current sources.

Analyzing a circuit using mesh or nodal analysis usually requires solving several linear equations. The superposition theorem allows us to simplify the analysis of a multi-source circuit by considering only one source at a time.

The superposition theorem states:

The voltage across (or the current through) a resistor may be determined by finding the sum of the effects due to each independent source in the circuit.

In order to determine the effects due to one source, it is necessary to remove all other sources from the circuit. This is accomplished by replacing voltage sources with short circuits and by replacing current sources with open circuits.

CALCULATIONS

1. Use superposition to calculate currents I_1, I_2, and I_3 in the circuit of Figure 6-1. Record the results in Table 6-2.

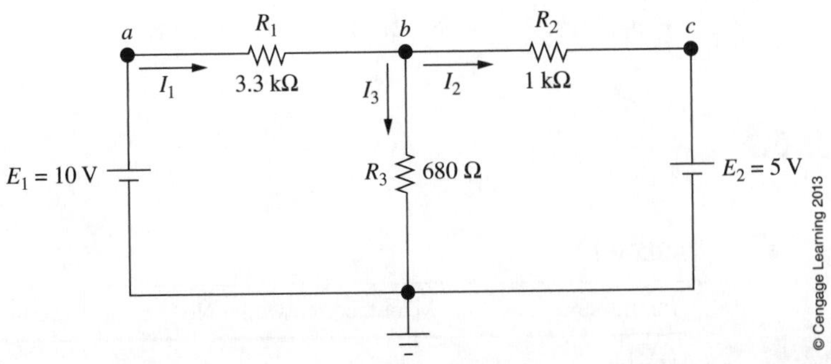

FIGURE 6-1

TABLE 6-2

I_1	
I_2	
I_3	

© Cengage Learning 2013

2. Use superposition to calculate the currents I_1, I_2, and I_3 in the circuit of Figure 6-2. Record the results in Table 6-3.

TABLE 6-3

I_1	
I_2	
I_3	

© Cengage Learning 2013

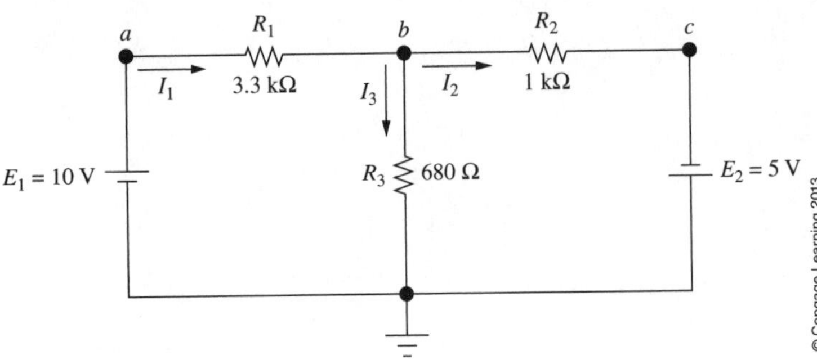

FIGURE 6-2

MEASUREMENTS

3. Assemble the circuit of Figure 6-1. Measure the voltages V_{ab}, V_b, and V_{bc}. Record your results in Table 6-4, showing the correct polarity for each measurement. Use these measurements and the resistor color codes to calculate the currents I_1, I_2, and I_3 for the circuit. Show the correct polarity for each current. (Use a negative sign to indicate that current is opposite to the indicated direction.)

TABLE 6-4

V_{ab}	
V_b	
V_{bc}	
I_1	
I_2	
I_3	

© Cengage Learning 2013

4. Assemble the circuit of Figure 6-2. Measure the voltages V_{ab}, V_b, and V_{bc}. Record your results in Table 6-5, showing the correct polarity for each measurement. Use these measurements and the resistor color codes to calculate the currents I_1, I_2, and I_3 for the circuit. Show the correct polarity for each current.

TABLE 6-5

V_{ab}	
V_b	
V_{bc}	
I_1	
I_2	
I_3	

5. Remove the voltage source E_2 from the circuit of Figure 6-1 and replace it with a short circuit. Measure the voltages V_{ab}, V_b, and V_{bc} and calculate the currents I_1, I_2, and I_3 due to the voltage source E_1. Record these results in Table 6-6.

TABLE 6-6

$V_{ab(1)}$	
$V_{b(1)}$	
$V_{bc(1)}$	
$I_{1(1)}$	
$I_{2(1)}$	
$I_{3(1)}$	

6. Remove voltage source E_1 from the circuit of Figure 6-1 and replace it with a short circuit. Measure voltages V_{ab}, V_b, and V_{bc} and use the results to calculate currents I_1, I_2, and I_3 due to the voltage source E_2. Record the results in Table 6-7.

TABLE 6-7

$V_{ab(2)}$	
$V_{b(2)}$	
$V_{bc(2)}$	
$I_{1(2)}$	
$I_{2(2)}$	
$I_{3(2)}$	

7. Reverse the polarity of voltage source E_2 as shown in the circuit of Figure 6-2. (Voltage source E_1 is still substituted by a short circuit.) Measure voltages V_{ab}, V_b, and V_{bc} and use the results to calculate currents I_1, I_2, and I_3 due to the voltage source E_2. Record the results in Table 6-8.

TABLE 6-8

$V_{ab(3)}$	
$V_{b(3)}$	
$V_{bc(3)}$	
$I_{1(3)}$	
$I_{2(3)}$	
$I_{3(3)}$	

CONCLUSIONS

8. Compare the measured currents of Table 6-4 to the theoretical currents calculated in Step 1. Determine the percent deviation for each of the currents.

I_1(theoretical) = __________ percent variation = __________

I_2(theoretical) = __________ percent variation = __________

I_3(theoretical) = __________ percent variation = __________

9. Compare the measured currents recorded in Table 6-5 to the theoretical currents calculated in Step 2. Determine the percent deviation for each of the currents.

I_1(theoretical) = __________ percent variation = __________

I_2(theoretical) = __________ percent variation = __________

I_3(theoretical) = __________ percent variation = __________

10. Combine the results of Table 6-6 and Table 6-7.

$V_{ab} = V_{ab(1)} + V_{ab(2)} =$ __________

$V_b = V_{b(1)} + V_{b(2)} =$ __________

$V_{bc} = V_{bc(1)} + V_{bc(2)} =$ __________

$I_1 = I_{1(1)} + I_{1(2)} =$ __________

$I_2 = I_{2(1)} + I_{2(2)} =$ __________

$I_3 = I_{3(1)} + I_{3(2)} =$ __________

Compare the previous results to the measurements recorded in Table 6-4. According to superposition, the results should be the same.

____

____

11. Combine the results of Table 6-6 and Table 6-8.

$$V_{ab} = V_{ab(1)} + V_{ab(3)} = \_\_\_\_\_\_\_\_$$

$$V_b = V_{b(1)} + V_{b(3)} = \_\_\_\_\_\_\_\_$$

$$V_{bc} = V_{bc(1)} + V_{bc(3)} = \_\_\_\_\_\_\_\_$$

$$I_1 = I_{1(1)} + I_{1(3)} = \_\_\_\_\_\_\_\_$$

$$I_2 = I_{2(1)} + I_{2(3)} = \_\_\_\_\_\_\_\_$$

$$I_3 = I_{3(1)} + I_{3(3)} = \_\_\_\_\_\_\_\_$$

Compare the above results to the measurements recorded in Table 6-5. According to superposition, the results should be the same.

____

____

PROBLEMS

12. Apply mesh analysis to calculate the loop currents in the circuit of Figure 6-1. Use the loop currents to determine currents I_1, I_2, and I_3.

$I_{loop\ 1}$ = ____________

$I_{loop\ 2}$ = ____________

I_1	
I_2	
I_3	

13. Apply mesh analysis to calculate the loop currents in the circuit of Figure 6-2. Use the loop currents to determine currents I_1, I_2, and I_3.

$I_{loop\ 1}$ = ____________

$I_{loop\ 2}$ = ____________

I_1	
I_2	
I_3	

14. Use nodal analysis to calculate the node voltage V_b in the circuit of Figure 6-1. Your result should be very close to the measured value recorded in Table 6-4.

V_b = ____________

15. Use nodal analysis to calculate the node voltage V_b in the circuit of Figure 6-2. Your result should be very close to the measured value recorded in Table 6-5.

V_b = ____________

Name ______________________

Date ______________________

Class ______________________

Thévenin's and Norton's Theorems (dc)

OBJECTIVES

After completing this lab, you will be able to

- determine the Thévenin and Norton equivalent of a complex circuit,
- analyze a circuit using the Thévenin or Norton equivalent circuit,
- determine the value of load resistance needed to ensure maximum power transfer to the load,
- measure the Thévenin (or Norton) resistance of a circuit using an ohmmeter,
- measure the Thévenin voltage of a circuit using a voltmeter,
- measure the Norton current of a circuit using an ammeter,
- describe how maximum power is transferred to a load when the load resistance is equal to the Thévenin resistance ($R_L = R_{Th}$).

EQUIPMENT REQUIRED

☐ Digital multimeter (DMM)
☐ dc power supply
Note: Record this equipment in Table 7-1.

COMPONENTS

☐ Resistors: 3.3-kΩ, 4.7-kΩ, 5.6-kΩ (1/4-W, 5% tolerance)
10-kΩ variable resistor

EQUIPMENT USED

TABLE 7-1

Instrument	Manufacturer/Model No.	Serial No.
DMM		
dc Supply		

TEXT REFERENCE

Section 9.2 THÉVENIN'S THEOREM
Section 9.3 NORTON'S THEOREM
Section 9.4 MAXIMUM POWER TRANSFER THEOREM

DISCUSSION

Thévenin's theorem:

Any linear bilateral network may be reduced to a simplified two-terminal network consisting of a single voltage source, E_{Th}, in series with a single resistor, R_{Th}. Once the original network is simplified, any load connected to the output terminals will behave exactly as if the load were connected in series with E_{Th} and R_{Th}.

Norton's theorem:

Any linear bilateral network may be reduced to a simplified two-terminal network consisting of a single current source, I_N, in parallel with a single resistor, R_N. A Thévenin equivalent circuit is easily converted into a Norton equivalent by performing a source conversion as follows:

$$R_N = R_{Th} \tag{7-1}$$

$$I_N = \frac{E_{Th}}{R_{Th}} \tag{7-2}$$

When a load is connected across the output terminals, the circuit will behave exactly as if the load were connected in parallel with I_N and R_N.

Maximum power transfer theorem:

Maximum power will be delivered to the load resistance when the load resistance is equal to the Thévenin (or Norton) resistance.

CALCULATIONS

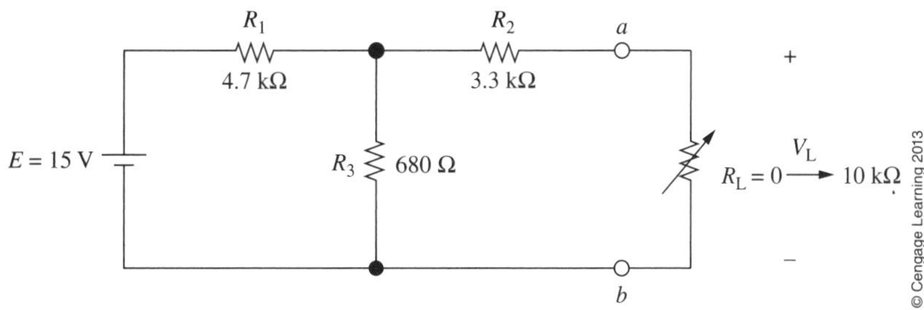

FIGURE 7-1

1. Determine the Thévenin equivalent of the circuit of Figure 7-1. Sketch the equivalent circuit in the space provided below.

2. Determine the Norton equivalent of the circuit of Figure 7-1. Sketch the equivalent circuit in the space provided below.

3. Calculate the minimum and the maximum voltage V_L which will appear across the load as R_L is varied between 0 and 10 kΩ. Enter the results in Table 7-1.

TABLE 7-1

$V_{L(min)}$	
$V_{L(max)}$	

© Cengage Learning 2013

4. Calculate the minimum and the maximum load current I_L which will occur through the load as R_L is varied between 0 and 10 kΩ. Enter the results in Table 7-2.

TABLE 7-2

$I_{L(min)}$	
$I_{L(max)}$	

© Cengage Learning 2013

5. Determine the value of load resistance for which maximum power will be transferred to the load.

MEASUREMENTS

6. Assemble the circuit of Figure 7-1, temporarily omitting the load resistor R_L. Insert the DMM voltmeter across terminals *a* and *b* and measure the open-circuit voltage. This is the Thévenin voltage E_{Th}. Record the measurement here.

E_{Th}	

7. Insert the DMM ammeter between terminals *a* and *b*. Ensure that the ammeter is adjusted to measure the expected Norton current. Because the ammeter is effectively a short circuit, you are measuring the short-circuit current between terminals *a* and *b*. Record the measurement here.

I_N	

8. Remove the voltage source from the circuit and replace it with a short circuit. Place the DMM ohmmeter across terminals *a* and *b* and measure

the resistance between these terminals. This is the Thévenin resistance R_{Th}. Record the result here.

R_{Th}	

9. Reconnect the voltage source into the circuit. Connect the variable 10-kΩ resistor as a rheostat and insert it as the load resistance R_L. Adjust the variable resistor between its minimum and maximum values. Measure and record the maximum and minimum output voltage V_L in Table 7-3.

TABLE 7-3

$V_{L(min)}$	
$V_{L(max)}$	

© Cengage Learning 2013

10. Place the DMM ammeter in series with the load resistor R_L. Adjust the variable resistor between its minimum and maximum values. Measure and record the minimum and maximum load current I_L in Table 7-4.

TABLE 7-4

$I_{L(min)}$	
$I_{L(max)}$	

© Cengage Learning 2013

11. Connect the DMM voltmeter across the load resistor. Adjust R_L until the output voltage is exactly half of the Thévenin voltage measured in Step 6. When $V_L = E_{Th}/2$, the load resistor is receiving the maximum amount of power from the circuit. Carefully remove R_L from the circuit, ensuring that the rheostat is not accidently readjusted. Use the DMM ohmmeter to measure the value of R_L. Record the measurement here.

R_L	

CONCLUSIONS

12. Compare the measured value of Thévenin voltage E_{Th} in Step 6 to the calculated value of Step 1. Determine the percent variation.

$E_{Th(theoretical)}$ = ____________ percent variation = ____________

13. Compare the measured value of Thévenin (or Norton) resistance R_{Th} in Step 8 to the calculated value of Step 1. Determine the percent variation.

$R_{Th(theoretical)}$ = ____________ percent variation = ____________

14. Compare the measured value of Norton current I_N in Step 7 to the calculated value of Step 2. Determine the percent variation.

$I_{N(theoretical)} =$ ____________ percent variation = ____________

15. An alternate method of determining the Thévenin (Norton) resistance is by applying Ohm's law to the Thévenin voltage and Norton current. Calculate the value of Thévenin resistance using the measured values of Thévenin voltage and Norton current.

$$R_{Th} = R_N = \frac{E_{Th}}{I_N} \tag{7-3}$$

R_{Th}	

Compare this value to the actual measured value of Thévenin resistance of Step 8.

____

____

16. Compare the measured minimum and maximum voltage V_L to the calculated voltages determined in Step 3. Explain why there is a slight variation.

____

____

17. Compare the measured minimum and maximum load current I_L to the calculated currents determined in Step 4.

____

18. When delivering maximum power to the load, how did the actual value of load resistance R_L compare to the theoretical value calculated in Step 5?

____

____

Name ______________________

Date ______________________

Class ______________________

LAB 8

Capacitors

OBJECTIVES

After completing this lab, you will be able to
- measure capacitance,
- verify capacitor relationships for series and parallel connections,
- verify that a capacitor behaves as an open circuit for steady state dc.

EQUIPMENT REQUIRED

- ☐ Digital multimeter (DMM)
- ☐ Capacitance meter
- ☐ Variable dc power supply

COMPONENTS

☐ Capacitors: One each of 1 μF, 0.47 μF, and 0.33 μF, nonelectrolytic, 35 WVDC or greater

☐ Resistors: One each of 2.7-kΩ, 3.9-kΩ, and 10-kΩ, 1/4 W

EQUIPMENT USED

TABLE 8-1

Instrument	Manufacturer/Model No.	Serial No.
DMM		
Capacitance meter*		
Power supply		

*Such as a DMM with capacitance measuring capability, an LCR meter, a digital capacitance bridge, or an impedance bridge.

TEXT REFERENCE

Section 10.1 CAPACITANCE
Section 10.7 CAPACITORS IN PARALLEL AND SERIES

DISCUSSION

Practical Note

The farad is a very large quantity; practical capacitors generally range in value from a few picofarads to a few hundred microfarads.

A *capacitor* is a charge storage device, and its electrical property is called *capacitance*. The more charge that a capacitor can store for a given voltage, the larger its capacitance. Capacitance, charge, and voltage are related by the equation

$$C = \frac{Q}{V} \qquad (8\text{-}1)$$

where C is capacitance in *farads,* Q is charge stored (in coulombs), and V is the capacitor terminal voltage (in volts). Because of its ability to store charge, a capacitor holds its voltage. That is, if you charge a capacitor, then disconnect the source, a voltage will remain on the capacitor for a considerable length of time. Dangerous voltages can be present on charged capacitors. For this reason, you should discharge capacitors before working with them.

For capacitors in parallel, the total capacitance is the sum of the individual capacitances. That is,

$$C_T = C_1 + C_2 + ... + C_N \qquad (8\text{-}2)$$

For capacitors in series, the total capacitance may be found from

$$\frac{1}{C_T} = \frac{1}{C_1} + \frac{1}{C_2} + \ldots + \frac{1}{C_N} \qquad (8\text{-}3)$$

Steady State Capacitor Currents and Voltages

Since capacitors consist of conducting plates separated by an insulator, there is no conductive path from terminal to terminal. Thus, when a capacitor is placed across a dc source, its steady state current is zero. This means that a capacitor in steady state looks like an open circuit to dc.

When connected in parallel, the voltage across all capacitors is the same. However, when connected in series, voltage divides in inverse proportion to the size of the capacitances: that is, the smaller the capacitance, the larger the voltage. For capacitors in series as in Figure 8-3(a), the steady dc voltages on capacitors are related by

$$V_x = \left(\frac{C_T}{C_x}\right)E, \quad V_1 = \left(\frac{C_2}{C_1}\right)V_2, \quad V_1 = \left(\frac{C_3}{C_1}\right)V_3 \tag{8-4}$$

and so on. (This is the voltage divider rule for capacitance.)

Measuring Capacitance

Measuring capacitance with a modern capacitor tester is straightforward. The general procedure is

1. Short the capacitor's leads to discharge the capacitor. Remove the short.
2. Set the function selector of the tester to the appropriate capacitance range (if not autoranging), then connect the capacitor. Observe polarity markings on polarized capacitors if applicable. (This is not necessary on some testers.)
3. Read the capacitance value directly from the numeric readout.

With such testers, it takes only a few seconds to measure capacitance.

MEASUREMENTS

1. Carefully measure each capacitor and resistor and record their values in Tables 8-2 and 8-3.

TABLE 8-2

	Nominal	Measured
C_1	1 μF	
C_2	0.47 μF	
C_3	0.33 μF	

© Cengage Learning 2013

TABLE 8-3

	Nominal	Measured
R_1	2.7 KΩ	
R_2	3.9 KΩ	
R_3	10 KΩ	

© Cengage Learning 2013

2. a. Assemble the circuit of Figure 8-1(a), measure capacitance C_T, and record in Table 8-4. Repeat for circuits (b), (c), and (d).
   b. Verify the values measured in (a) by analyzing each circuit using the measured capacitor values from Table 8-2. Record calculated values in Table 8-4. How do they compare to the measured results?

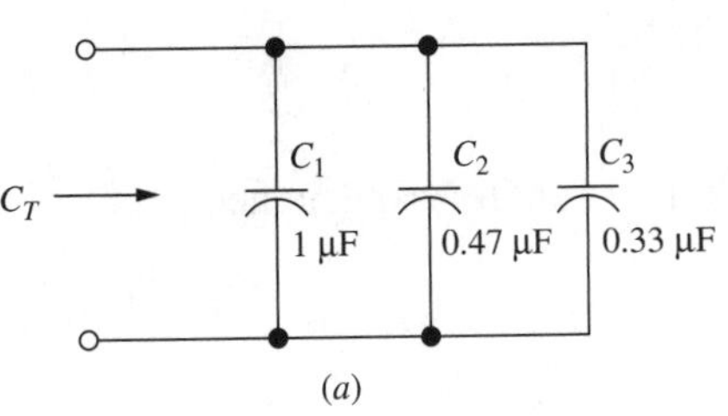

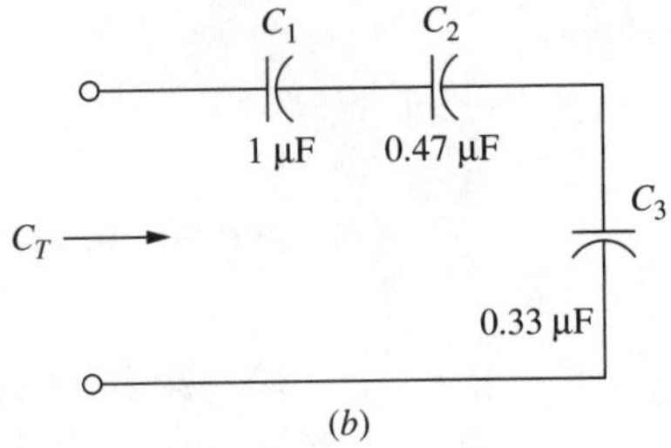

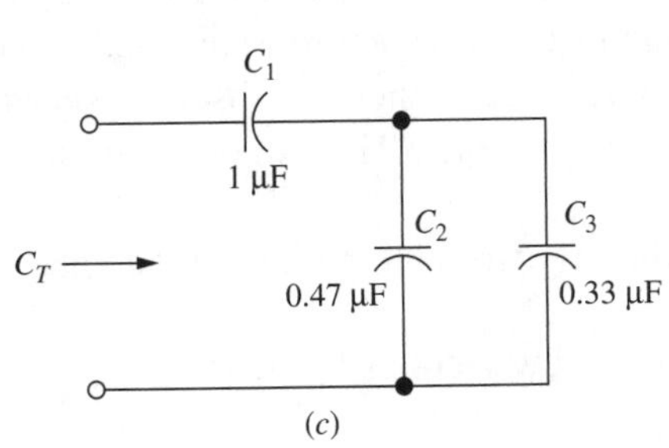

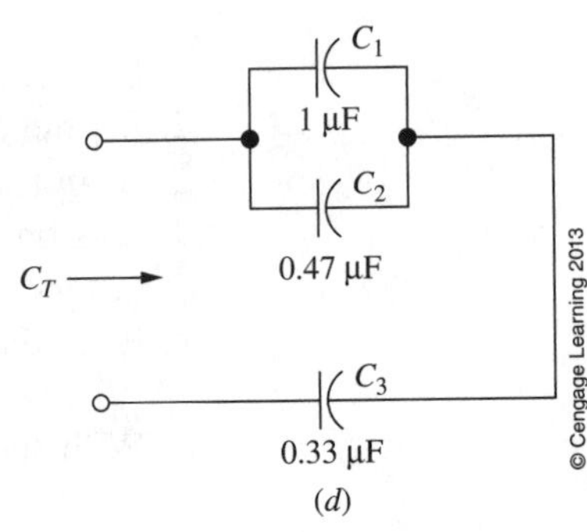

FIGURE 8-1 Circuits for Test 2.

TABLE 8-4

Circuit	Total Capacitance C_T (μF) Measured	Total Capacitance C_T (μF) Calculated	% of Difference
Figure 8-1(a)			
Figure 8-1(b)			
Figure 8-1(c)			
Figure 8-1(d)			

3. a. Assemble the circuit of Figure 8-2. Set $E = 18.0$ V, measure V_1, V_2, and V_3, and record in Table 8-5.

TABLE 8-5

	Measured	Computed
V_1		
V_2		
V_3		

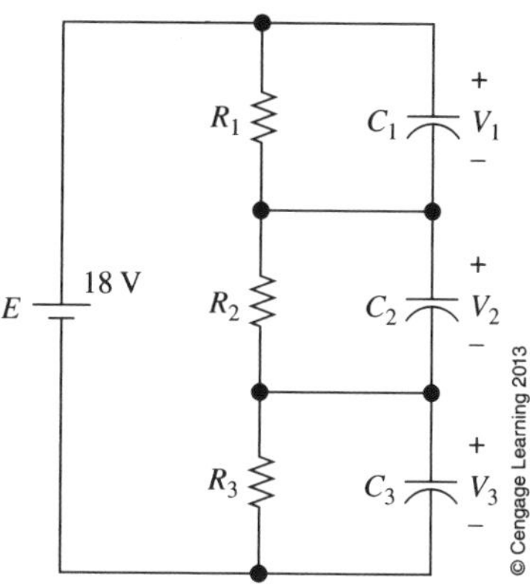

FIGURE 8-2 Circuits for Test 3.

b. Verify the results of Table 8-5 by analyzing the circuit of Figure 8-2 and calculating voltages. Why do the equations in Equation 8-4 not apply here? Assuming steady state dc, why do the capacitors not load the resistive circuit?

PROBLEMS

4. For the circuit of Figure 8-3(a), compute the voltage across each capacitor and record in Table 8-6. Repeat for Figure 8-3(b).

TABLE 8-6

	Figure 8-3(a)	Figure 8-3(b)
V_1		
V_2		
V_3		

5. For the circuit of Figure 8-4, compute the voltage across the capacitor.

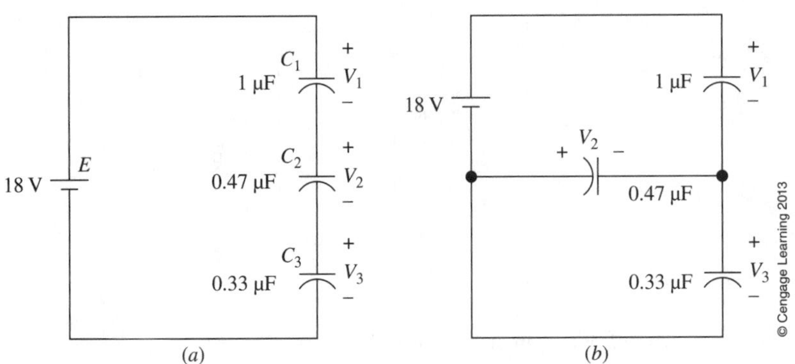

FIGURE 8-3 Circuits for Problem 4.

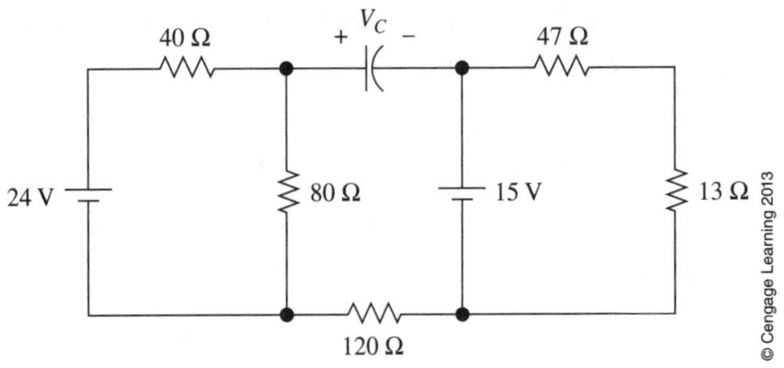

FIGURE 8-4 Circuits for Problem 5.

FOR FURTHER INVESTIGATION AND DISCUSSION

Capacitors can have quite loose tolerances—that is, the actual capacitance of a capacitor may differ by as much as ±10% to ±20% from its nominal value (i.e., its body marking) depending on the type of capacitor used. Consider two capacitors, $C_1 = 1\ \mu F \pm 10\%$ and $C_2 = 0.47\ \mu F \pm 20\%$. Write a short discussion paper on the impact of tolerances. As the basis for your discussion, (a) place the capacitors in parallel and compute the nominal capacitance C_T and the minimum and maximum value that C_T will have if both capacitors are at their worst-case tolerance limits. (b) Repeat if the capacitors are in series. (c) With this as background, show how tolerance errors can impact total capacitance of a more complex circuit. For this last part, consider Figure 8-1(c). Assume that C_3 has a tolerance of ±10%.

Name ______________________________

Date ______________________________

Class ______________________________

LAB 9

Capacitor Charging and Discharging

OBJECTIVES

After completing this lab, you will be able to

- measure capacitor charge and discharge times,
- confirm the voltage/current direction convention for capacitors,
- confirm the Thévenin method of analysis for capacitive charging and discharging,

EQUIPMENT REQUIRED

- ☐ Digital multimeter (DMM), VOM
- ☐ Variable dc power supply
- ☐ Function generator (optional)
- ☐ Oscilloscope (optional)

COMPONENTS

☐ Capacitors:	470-μF (electrolytic), 0.01-μF (nonelectrolytic)
☐ Resistors:	10-kΩ, 20-kΩ, 39-kΩ, 47-kΩ, 1/4-W
☐ Switch:	Single pole, double throw
☐ Stopwatch	

NOTE TO THE INSTRUCTOR

Part D of this lab may be run as an instructor demo if your students have not yet had instruction on the oscilloscope.

PRE-STUDY (OPTIONAL)

Interactive simulations *CircuitSim 11-1* (Test 1 only), *11-2,* and *11-17B* from the Student Premium Website may be used as pre-study for this lab. If you have not already done so, go to the Student Premium Website, select *Learning Circuit Analysis Interactively on Your Computer,* download and unzip *Ch 11,* and run simulations *Sim 11-1,* and so on.

EQUIPMENT USED

TABLE 9-1

Instrument	Manufacturer/Model No.	Serial No.
DMM		
Power supply		
Function generator (optional)		
Oscilloscope (optional)		

TEXT REFERENCE

Section 10.8 CAPACITOR CURRENT AND VOLTAGE
Section 11.1 INTRODUCTION
Section 11.2 CAPACITOR CHARGING EQUATIONS
Section 11.4 CAPACITOR DISCHARGING EQUATIONS
Section 11.5 MORE COMPLEX CIRCUITS
Section 11.8 TRANSIENT ANALYSIS USING COMPUTERS

DISCUSSION

Capacitor charging and discharging may be studied using the circuit of Figure 9-1. When the switch is in position 1, the capacitor charges at a rate determined by its capacitance and the resistance through which it charges; when the switch is in position 2, it discharges at a rate determined by its capacitance and the resistance through which it discharges. This phenomenon of charging and discharging is important as it affects the operation of many circuits.

Caution

In this lab, you use electrolytic capacitors. Electrolytics are polarized and must be used with their + lead connected to the positive side of the circuit and their − lead to the negative side. An incorrectly connected electrolytic may explode.

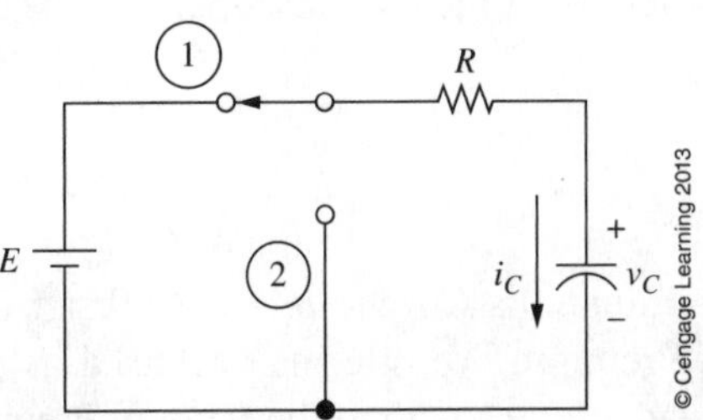

FIGURE 9-1 Circuit for studying capacitor charging and discharging.

Voltage and Current During Charging

Consider Figure 9-2. At the instant the switch is moved to the charge position, current jumps from zero to E/R (since the capacitor looks like a short circuit at this instant). As the capacitor voltage approaches full source voltage, current approaches zero (since the capacitor looks like an open circuit to dc). During charging, voltage and current are given by

$$i_c = \frac{E}{R} e^{-t/RC} \tag{9-1}$$

$$v_c = E(1 - e^{-t/RC}) \tag{9-2}$$

The product RC is referred to as the time constant and is given the symbol τ. Thus,

$$\tau = RC \tag{9-3}$$

In one time constant, the capacitor voltage climbs to 63.2% of its final value while the current drops to 36.8% of its initial value. For all practical purposes, charging is complete in five time constants.

Voltage and Current During Discharging

The discharge curves are shown in Figure 9-3. When the switch is moved to position 2, the capacitor looks momentarily like a voltage source with value

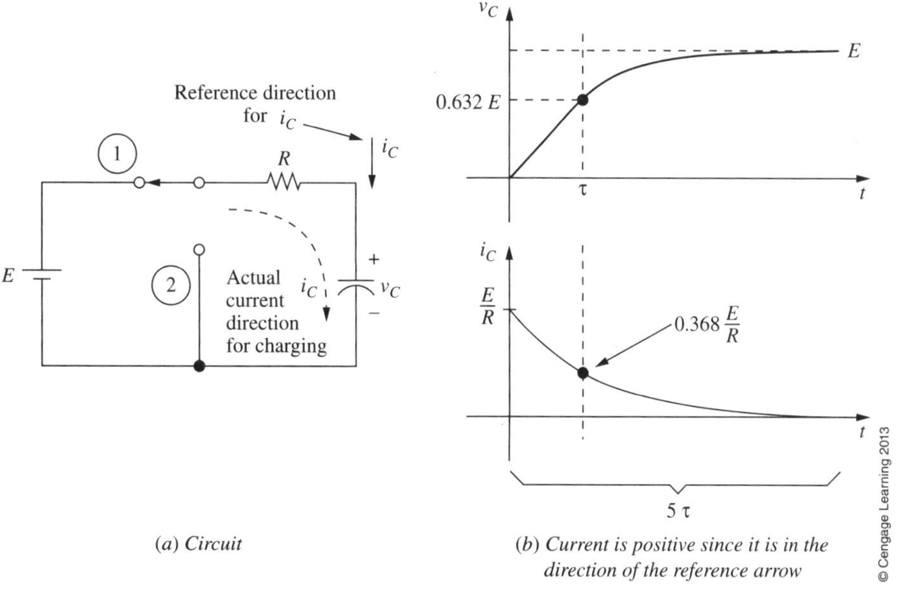

(*a*) *Circuit*

(*b*) *Current is positive since it is in the direction of the reference arrow*

FIGURE 9-2 Capacitor voltage and current during charging. Capacitor is initially uncharged.

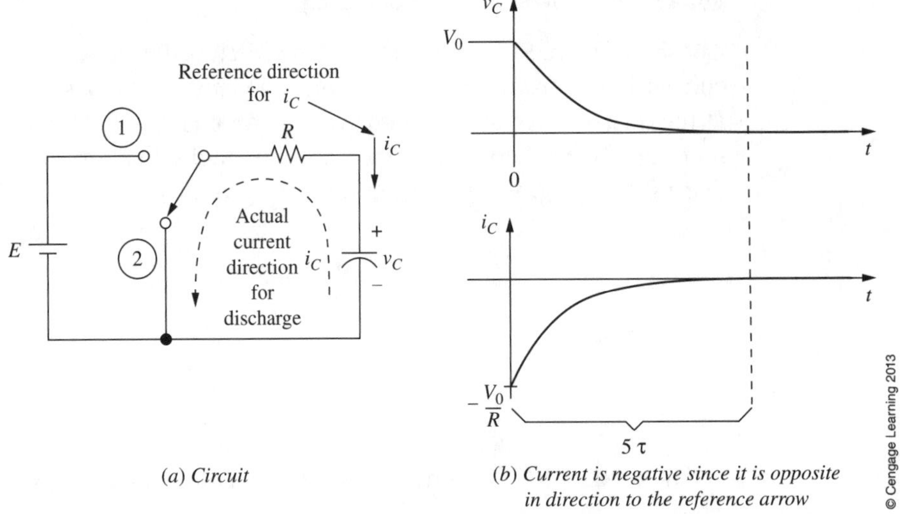

FIGURE 9-3 Capacitor voltage and current during discharging.

V_o where V_o is the voltage on the capacitor at the instant the switch is moved. (If the capacitor is fully charged, $V_o = E$.) Current jumps from zero to $-V_o/R$, then decays to zero. [It is negative since it is opposite in direction to the reference as indicated in Figure 9-3(a).] Voltage decays from V_o to zero. Equations are

$$i_C = -\frac{V_o}{R}e^{-t/RC} \quad \text{or} \quad i_c = -\frac{E}{R}e^{-t/RC} \tag{9-4}$$

$$v_C = V_o e^{-t/RC} \quad \text{or} \quad v_c = Ee^{-t/RC} \tag{9-5}$$

Discharging takes five time constants. Although for this particular circuit charge and discharge resistances are the same, in general they are different. For this latter case, charge and discharge time constants will be different.

A Final Note

To properly observe capacitor charging and discharging, you need an oscilloscope. Since you may not have studied the oscilloscope at this time, we will examine the basic ideas in other ways. However, at the end of the lab, we have included an instructor demonstration (Part D) using an oscilloscope. Here, you will be able to observe capacitor charging and discharging directly on the screen. (You can also study charging and discharging interactively on your computer by using the Multisim simulations noted at the start of the lab.)

MEASUREMENTS

Practical Notes

1. Use a VOM here with a sensitivity of at least 20,000 ohms per volt to minimize circuit loading.
2. Electrolytic capacitors are polarized devices that require a dc voltage to set up and maintain their oxide film. In this test, the capacitor's voltage varies as it charges and discharges, and thus its oxide film thickness may change throughout the test, resulting in a change in the effective capacitance of the capacitor. As a consequence, the results that you get may differ somewhat from what you calculate (even if you measure the capacitor's value first).

PART A: Charge and Discharge Times

For Part A, use a VOM to observe capacitor voltage and a stopwatch to time charging and discharging. For the large capacitance value required here, you need an electrolytic capacitor. Unfortunately, you will find that your results agree only moderately well with theory—see boxed Practical Note 2. However, the results clearly verify the theory.

1. Assemble the circuit of Figure 9-1 with $R = 10\ \text{k}\Omega$, $C = 470\ \mu\text{F}$, and source voltage about 25 V. Place the switch in the charge position and wait for the capacitor to fully charge. Carefully adjust E until $V = 25$ V across the capacitor.

   a. Compute the time constant for the circuit.

   $\tau_{\text{computed}} = RC =$ ____________.

   b. Return the switch to the discharge position. Wait for the capacitor to fully discharge, then move the switch to the charge position and using the stopwatch, time how long it takes for the capacitor voltage to reach 15.8 V (i.e., 63.2% of its final voltage). This is the measured time constant.

   $\tau_{\text{measured}} =$ ____________ (charging).

   c. Hold the switch in the charge position until the capacitor voltage stabilizes at 25 V. Now move the switch to the discharge position and time how long it takes for the voltage to drop to 9.2 V (i.e., to 36.8% of its initial value).

   $\tau_{\text{measured}} =$ ____________ (discharging).

2. Change R to 20 kΩ and repeat Test 1.

   a. $\tau_{\text{computed}} =$ ____________

   b. $\tau_{\text{measured}} =$ ____________ (charging)

   c. $\tau_{\text{measured}} =$ ____________ (discharging)

3. Describe how the results of Tests 1 and 2 confirm the theory.

____

____

Note

Since the actual discharge current direction of Figure 9-3(a) is counterclockwise, you might be tempted to reverse the current reference arrow for i_C. Do not do this, as it leads to mathematically inconsistent results. For example, the minus sign in Equation 9-4 means that the actual current is opposite in direction to your chosen reference direction for i_C, and if you were to reverse the direction of i_C but keep the minus sign, this would say that the actual discharge current is in the clockwise direction, which it clearly is not. Additionally, if you change the direction of i_C, then the relationship

$$i_C = C\frac{dv_C}{dt}$$

yields the incorrect sign for current. (These statements are easily verified with Multisim.)

PART B: Current Direction

To conform to the standard voltage/current convention, the + sign for voltage v_C must be at the tail of the current direction arrow i_C as indicated in Figure 9-1. This is obviously correct for charging as indicated in Figure 9-2. During discharge, the polarity of the voltage does not change. Thus, for discharging, the current direction arrow must also remain in the clockwise direction, even though we know that actual current is in the opposite direction, as indicated in Figure 9-3. The interpretation is that for charging, current is in the same direction as the reference and hence, is positive, while for discharging, it is opposite to the reference and hence, is negative—see Note.

4. Add a DMM with autopolarity as in Figure 9-4. Connect the current input jack A to the positive side of the circuit as indicated. (Since the meter measures current *into* terminal A, this connection will yield a positive value when current is in the reference direction and a negative value when opposite to the reference.) We now verify charge and discharge directions experimentally.

   a. Move the switch to charge and note the sign of the multimeter reading. (Do not try to read its value—we are only interested in its sign.) Sign __________. Thus, the actual direction of current is

   ____.

   b. Move the switch to discharge and again note the sign. Sign __________. Thus, the actual direction of current is

   ____.

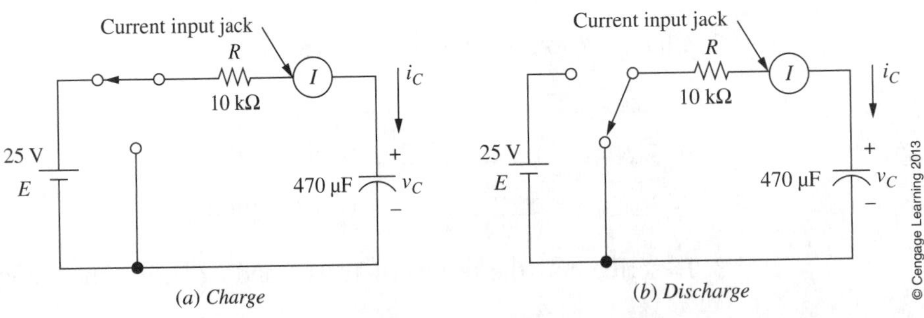

FIGURE 9-4 Verifying the current direction convention.

PART C: More Complex Circuits

For analysis, complex circuits can be reduced to their Thévenin equivalent. In this test, you will verify the Thévenin equivalent method.

5. a. Using Thévenin's theorem, determine the charge and discharge equivalent circuits for the circuit of Figure 9-5(a). Sketch in (b) and (c) respectively.

   b. From the Thévenin equivalents, compute the charge and discharge time constants.

   τ_{charge} = ____________. $\tau_{discharge}$ = ____________.

   c. For charging, compute capacitor voltage at $t = \tau_{charge}$.

   v_C = ____________.

   d. Assume the switch has been in the charge position long enough for the capacitor to fully charge. Now move the switch to discharge and compute the voltage after one discharge time constant. v_C = ____________.

   e. Assemble the circuit of Figure 9-5(a). Move the switch to charge and with the stopwatch, measure how long it takes for the voltage to reach the value determined in Test 5(c).

   $\tau_{charge(measured)}$ = ____________.

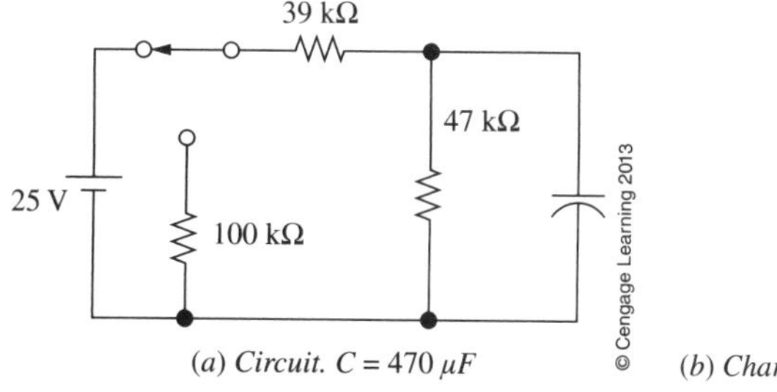

(a) Circuit. $C = 470\ \mu F$ (b) Charge equivalent (c) Discharge equivalent

FIGURE 9-5 Thévenin analysis.

f. Allow the capacitor to fully charge, then move the switch to discharge and with the stopwatch, measure the time that it takes for the voltage to reach the value determined in Test 5(d).

$\tau_{\text{discharge(measured)}} =$ ______________.

Discuss results. In particular, how well do the Thévenin equivalents represent the charging/discharging behavior of the capacitor in the real circuit?

Preliminary Note

Part D is set up as an instructor demonstration. However, if you have already learned how to use the oscilloscope, you may perform this part yourself.

PART D: Charging/Discharging Waveforms

The circuit is shown in Figure 9-6. A function generator simulates switching by applying a signal that cycles between voltage E and ground. A high rate of switching and a small time constant are used to give a waveform that can be viewed on the oscilloscope. Since a nonelectrolytic capacitor is used, agreement between theory and practice will be much better than in previous tests.

6. a. Measure R and C, then assemble the circuit of Figure 9-6. Using the function generator, apply an input with 0.5 ms high and low times (i.e., $f = 1$ kHz) and adjust the input to 5 V as indicated in Figure 9-6. Set the oscilloscope to display only the charging voltage. Sketch as Figure 9-7.

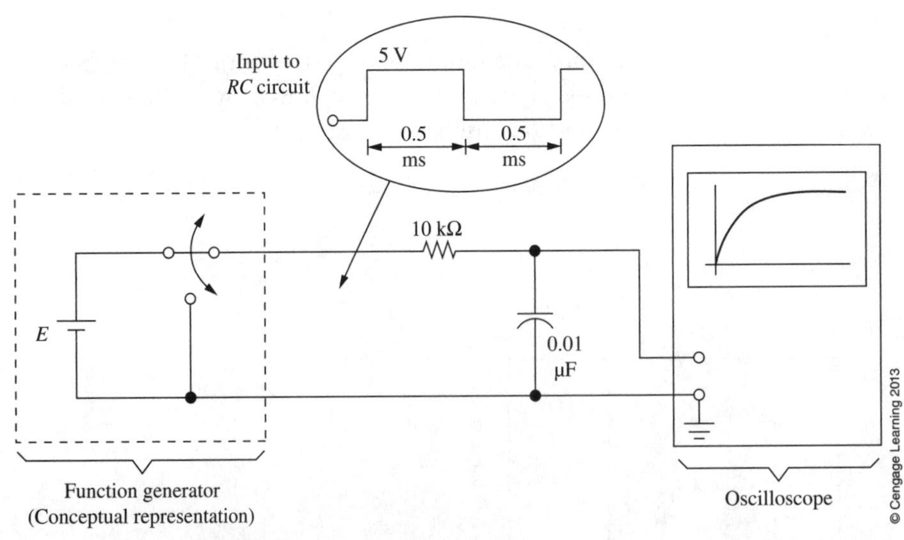

FIGURE 9-6 Test setup for Part D.

FIGURE 9-7 Waveform for Test 6.

b. Calculate the time constant using the measured R and C.

$\tau =$ ____________.

c. From the scope screen, measure capacitor voltages at $t = \tau, 2\tau, 3\tau, 4\tau,$ and 5τ, and tabulate.
d. Using circuit analysis techniques, compute voltages at these points. How well do they agree?
e. Display the discharge portion of the waveform and measure capacitor voltages at $t = \tau, 2\tau, 3\tau, 4\tau,$ and 5τ, and tabulate.
f. Repeat part (d) for the discharge values.

COMPUTER ANALYSIS

PSpice Users

Set V1, V2, PW, and PER, TR, TF, and TD to 0V, 5V, 0.5ms, 1ms, 1ns, 1ns, and 0, respectively. This creates the desired waveform.

7. Using either Multisim or PSpice, set up the *RC* circuit of Figure 9-6 with a pulse source that supplies 5 V for 0.5 ms and 0 V for the next 0.5 ms as indicated in Figure 9-6. (See Section 11.8 or interactive simulation *Sim 11-17B* of the textbook for reference. Do not forget to use a ground on the bottom end of the source symbol.) Set the capacitor initial voltage to zero and run a transient analysis. Using the cursor, determine voltages at $t = \tau, 2\tau, 3\tau, 4\tau,$ and 5τ for both the charge and discharge cases. Compare the results to those obtained from the oscilloscope, Test 6.

PROBLEMS

8. Assume the circuit of Figure 9-1 has a time constant of 100 μs. If R is doubled and C is tripled, calculate the new time constant.

9. For Figure 9-1, replace the wire between switch position 2 and common with a resistor R_2. If $R_2 = 4\,R$ and the capacitor takes 25 ms to reach full charge, how long will it take to discharge?

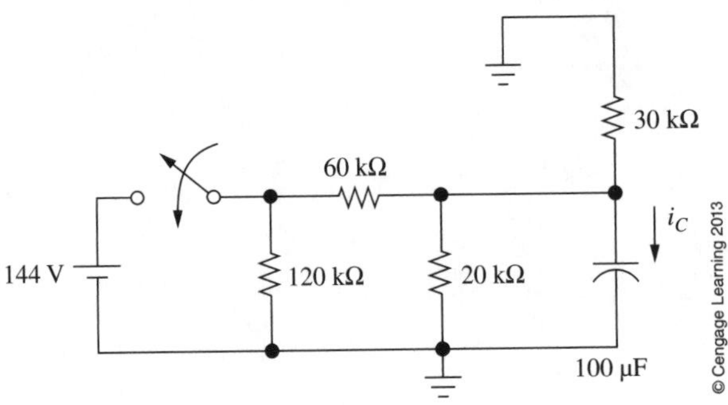

FIGURE 9-8 Circuit for Problem 10.

10. For Figure 9-8, the switch is closed at $t = 0$ s and opened 5 s later. The capacitor is initially uncharged.
    a. Determine the capacitor current i_C at $t = 2$ s.
    b. Determine the capacitor current i_C at $t = 7$ s.

FOR FURTHER INVESTIGATION AND DISCUSSION

Tolerances on resistors and capacitors affect the time constant of circuits and thus affect circuits that rely on *RC* charging and discharging for their operation. To investigate, assume the resistor of Figure 9-6 has a tolerance of ±5% and the capacitor ±10%. Write a short discussion paper on the impact of these tolerances on rise and fall waveforms. In your discussion, determine the minimum and maximum values that τ may have, then plot the waveforms for these two cases as well as for the nominal case on the same graph. Using PSpice or Multisim, verify these waveforms. Discuss the ramifications of what you have learned.

Name ______________________

Date ______________________

Class ______________________

LAB 10

Inductors in dc Circuits

OBJECTIVES

After completing this lab, you will be able to
- determine steady state dc voltages and currents in an *RL* circuit,
- measure the inductive "kick" voltage in an *RL* circuit,
- measure the time constant of an *RL* circuit.

EQUIPMENT REQUIRED

- ☐ Digital multimeter (DMM)
- ☐ Variable dc power supply
- ☐ Function generator
- ☐ Oscilloscope

COMPONENTS

☐ Resistors:	10-Ω, 82-Ω (two), 1-kΩ (1/4-W), 47-Ω, 220-Ω (1/2-W), 120-Ω (two, each 2-W)
☐ Inductors:	One, 2.4-mH, powdered iron core, Hammond Part #1534 or equivalent; one approximately 1.5-H, iron core

NOTE TO THE INSTRUCTOR

Parts C and D of this lab may be run as instructor demos if your students have not yet had instruction on the oscilloscope.

PRE-STUDY (OPTIONAL)

Interactive simulations *CircuitSim 14-1* (Test 2) and *14-8* may be used as pre-study. From the Student Premium Website, select *Learning Circuit Analysis Interactively on Your Computer*, download and unzip *Ch 14*, and run the corresponding *Sim* files.

EQUIPMENT USED

TABLE 10-1

Instrument	Manufacturer/Model No.	Serial No.
DMM		
Power supply		
Function generator		
Oscilloscope		

TEXT REFERENCE

Section 13.2 INDUCED VOLTAGE AND INDUCTANCE
Section 13.7 INDUCTANCE AND STEADY STATE dc
Section 14.1 INTRODUCTION
Section 14.2 CURRENT BUILDUP TRANSIENTS
Section 14.4 DE-ENERGIZING TRANSIENTS

DISCUSSION

Induced voltage is determined by Faraday's law. For a coil with a nonmagnetic core, induced voltage is directly proportional to the rate of change of current and is given by

$$v_L = L\frac{di}{dt} \tag{10-1}$$

where L is the inductance of the coil and di/dt is the rate of change of the current in the coil. The induced voltage opposes the change in current.

For inductances in series, total inductance is the sum of individual inductances. Thus,

$$L_T = L_1 + L_2 + \ldots + L_N \tag{10-2}$$

For inductances in parallel, total inductance may be found from

$$\frac{1}{L_T} = \frac{1}{L_1} + \frac{1}{L_2} + \ldots + \frac{1}{L_N} \tag{10-3}$$

Since real inductors have coil resistance, Equation 10-3 can only be used in practice if coil resistance is negligible.

Steady State and Transient Response

Steady State dc: As Equation 10-1 shows, voltage results only when current changes. Thus if current is constant (as in steady state dc), the voltage across an inductance is zero. Consequently, *to steady state dc, an inductance looks like a short circuit.*

Current Buildup Transients: Simple current buildup transients in *RL* circuits may be studied using the circuit of Figure 10-1(a). When the switch is closed, current in the inductor is given by

$$i_L = \frac{E}{R_1}(1 - e^{-R_1 t/L}) \tag{10-4}$$

which has a final steady state value of E/R_1 amps as shown in (b). Voltage is given by

$$v_L = Ee^{-R_1 t/L} \tag{10-5}$$

As Figures 10-1(b) and (c) show, at the instant the switch is closed, current is zero, and full source voltage appears across the inductance. This means that *an inductor, with initial current of zero amps, looks like an open circuit.*

Current Decay Transients: Consider Figure 10-2(a). Assume that the current in the inductor at the instant the switch is opened is I_0. Since current cannot change instantaneously, *a current-carrying inductance looks momentarily like a current source of I_0 at the instant of switch operation.* Current then decays to zero according to

$$i_L = I_0 e^{-Rt/L} \tag{10-6}$$

where $R = R_1 + R_2$ for the circuit of Figure 10-2. Other voltages and currents can be obtained from these relationships using basic circuit principles. For example, the voltage v_2 across resistor R_2 is $-R_2\, i_L$. Thus, it has the same shape as i_L but is negative as indicated in (c). [Multiplying $-R_2$ times Equation 10-6 yields $v_2 = -I_0\, R_2\, e^{-Rt/L}$ as indicated in Figure 10-2(c).] Note that V_0 (which is equal to $-I_0\, R_2$), can be many times larger than the source voltage.

The time constant of an *RL* circuit is given by

$$\tau = L/R \tag{10-7}$$

where *R* is the resistance through which the inductor current builds or decays; for Figure 10-2, $\tau = L/R_1$ during current buildup and $\tau = L/(R_1 + R_2)$ during current decay. Steady state is reached in 5τ, where the appropriate τ (charge or discharge) must be used.

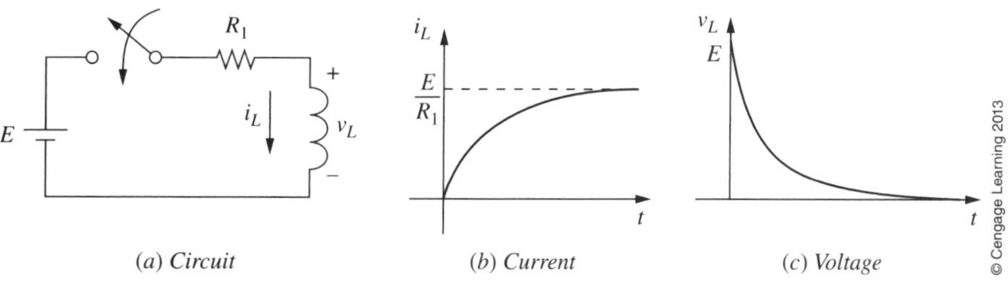

FIGURE 10-1 Current buildup transients.

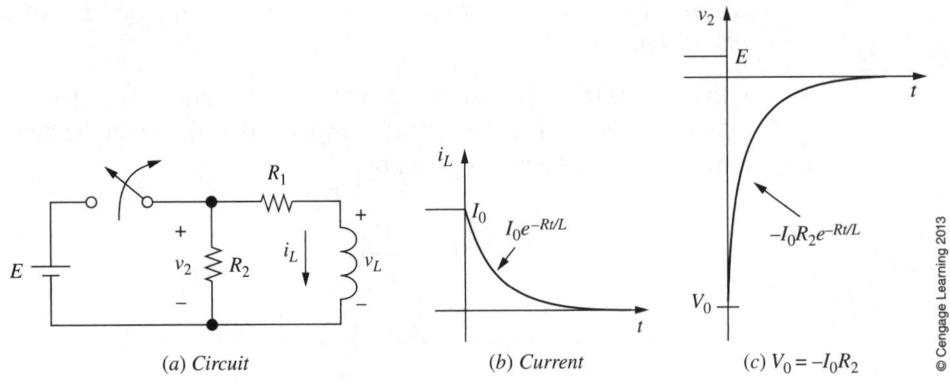

FIGURE 10-2 Current decay transients. $R = R_1 + R_2$.

MEASUREMENTS

PART A: Steady State Voltages and Currents

1. Measure the resistance of each 120-Ω resistor and the resistance of the 2.4-mH inductor and record in Table 10-2.
2. a. Assemble the *RL* portion of the circuit of Figure 10-3. With the source disconnected, measure input resistance R_{in}.

   $R_{in(measured)}$ = __________

   b. Using circuit analysis techniques, compute R_{in}. $R_{in(computed)}$ = _________. Compare to the measured value.

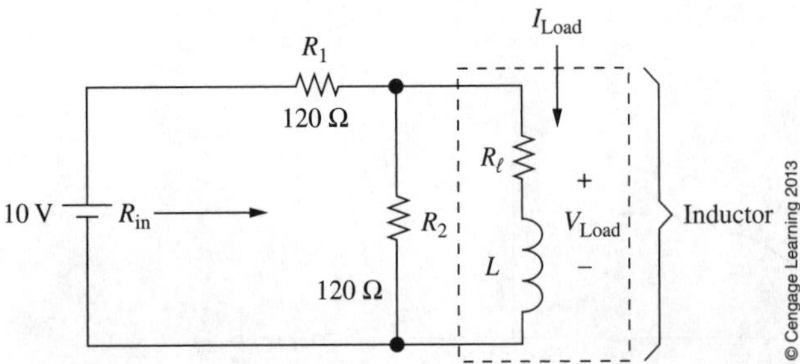

FIGURE 10-3 Circuit for Test 1.

TABLE 10-2

	Nominal	Measured
R_1	120 Ω	
R_2	120 Ω	
R_ℓ	Unknown	

© Cengage Learning 2013

c. Connect the source and measure voltage V_{Load} and I_{Load}.

$V_{\text{Load(measured)}} =$ ________________ $I_{\text{Load(measured)}} =$ ________________

d. Using circuit analysis techniques, solve for V_{Load} and I_{Load}. Compare to the measured values.

PART B: More Complex Steady State Circuits

For more complex circuits, use Thévenin's theorem to reduce portions of the circuit as necessary.

3. a. Measure each resistor for the circuit of Figure 10-4 and record in Table 10-3. Use the same inductor as in Figure 10-3.

   b. Determine the Thévenin equivalent of the circuit to the left of the inductor using circuit analysis techniques using the measured resistance values from Table 10-3.

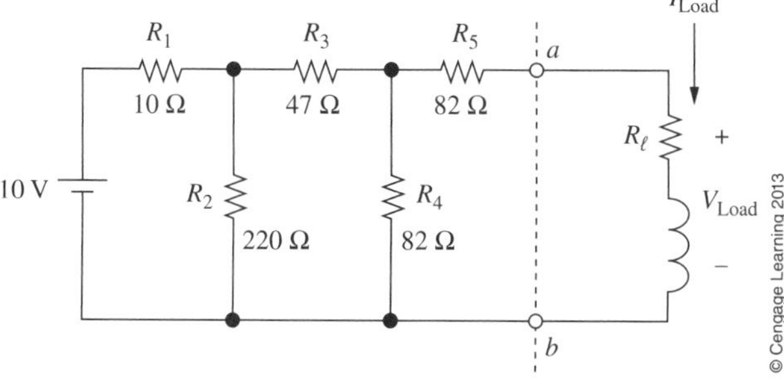

FIGURE 10-4 Circuit for Test 3. Use resistor values measured in Table 10-3.

TABLE 10-3

	Nominal	Measured
R_1	10 Ω	
R_2	220 Ω	
R_3	47 Ω	
R_4	82 Ω	
R_5	82 Ω	

c. Disconnect the inductor and measure the open circuit voltage across *a-b*.

(This is E_{Th}.) $E_{Th(measured)}$ = __________.

d. Disconnect the source and replace it with a short circuit. Measure the resistance looking back into the circuit. (This is R_{Th}.)

$R_{Th(measured)}$ = __________

e. Compare the measured and computed values for E_{Th} and R_{Th}.

f. Remove the short, reconnect the source and inductor, and measure I_{Load} and V_{Load}.

I_{Load} = __________ V_{Load} = __________

g. Using the Thévenin equivalent determined in (b), compute I_{Load} and V_{Load}. Compare to the measured values of (f).

Preliminary Note

Part C requires the use of an oscilloscope. If you have not yet covered the oscilloscope, this may be run as an instructor demonstration.

PART C: Inductive "Kick" Voltage

In this part of the lab, you will look at the *inductive kick* that results when current in an inductor is interrupted. Use the iron core inductor. (The inductor and resistor values used here are not critical. However, to see the results easily on the scope, you need a time constant of at least a few milliseconds. First determine appropriate resistor values.)

4. a. Measure the dc resistance of the inductor. R_ℓ = __________. Consider Figure 10-5(a). Define $R_1' = R_1 + R_\ell$. Select a value for R_1 such that current E/R_1' is easily handled by your power supply. Now select an R_2 that is about five or six times larger than R_1' and such that the discharge time constant $L/(R_1 + R_2 + R_\ell)$ is a few milliseconds or more. (If you know L, you can compute τ; if you have no way of measuring L, experiment until you get a waveform you can see.)

   b. Assemble the circuit with these values and connect the scope probe across resistor R_2 to view the inductive kick voltage v_2. [The waveform

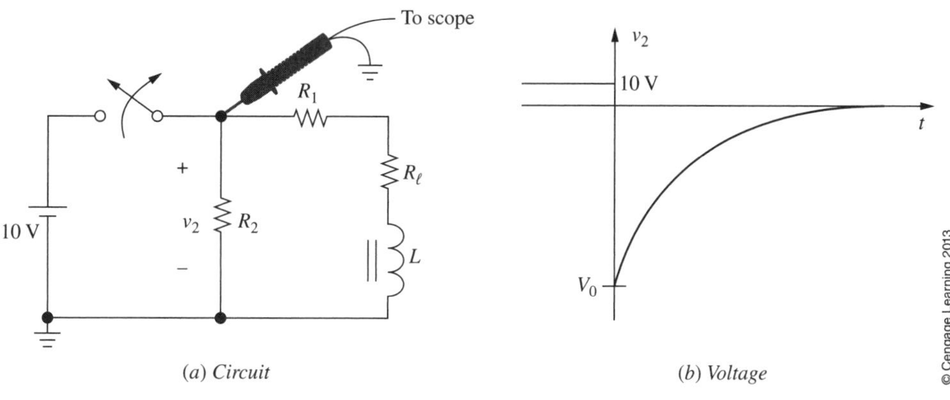

(a) Circuit (b) Voltage

FIGURE 10-5 Circuit for Test 4.

will look like that shown in Figure 10-5(b). Voltage V_0 depends on the ratio of R_2 to R_1'. If $R_2 = 5\,R_1'$ and $E = 10$ V, then $V_0 = -50$ volts.] The waveform is a single-shot event that is generated only when you open the switch. Set the scope triggering appropriately—for example, *Norm* or *Single Sweep*, negative slope. Depending on your scope, you may have difficulty in capturing a trace, but with some care (and perhaps, repeated operation of the switch), you should be able to observe the voltage spike of (b).

c. From the trace, measure the value of V_0. $V_0 =$ __________

d. Using the methods of Section 14.4 of the textbook, compute V_0 and compare to the measured value of (c). How do they compare?

PART D: Measuring the Time Constant

This part may be run as a demo if necessary.

5. a. Measure the resistance of the 1-kΩ resistor and the inductance of the 2.4-mH inductor. (Call the 1-kΩ resistor R_S.) If you do not have any way to measure inductance, use the nominal value marked on the inductor. Use R_ℓ from Table 10-2.

$R_\ell =$ __________ $L =$ __________ $R_S =$ __________

b. Assemble the circuit of Figure 10-6. The time constant for the circuit is $\tau = L/R_T$ where R_T is the sum of R_S, R_ℓ, and the output resistance R_{out} of your function generator. (A typical output resistance of a function generator is 50 Ω.) Thus,

$\tau =$ __________.

c. Set the function generator to the square wave mode at 40 kHz. (This provides ample time for current to build up and decay fully.) Adjust the signal generator and scope so that the amplitude of the v_S waveform is 5 grid lines (i.e., 5 V). The waveform should look like that shown in Figure 10-6.

d. Adjust the time base and triggering to get only the buildup waveform on the screen. Sketch this waveform as Figure 10-7.

e. Measure the time that it takes for the waveform to reach 63.2% of its final value. This is the measured time constant.

$\tau_{measured} =$ __________

Compare this value to the value computed in (b).

f. Change triggering to get the decay waveform on the screen. Measure the time constant here and compare to that of (d).

$\tau_{measured} =$ __________

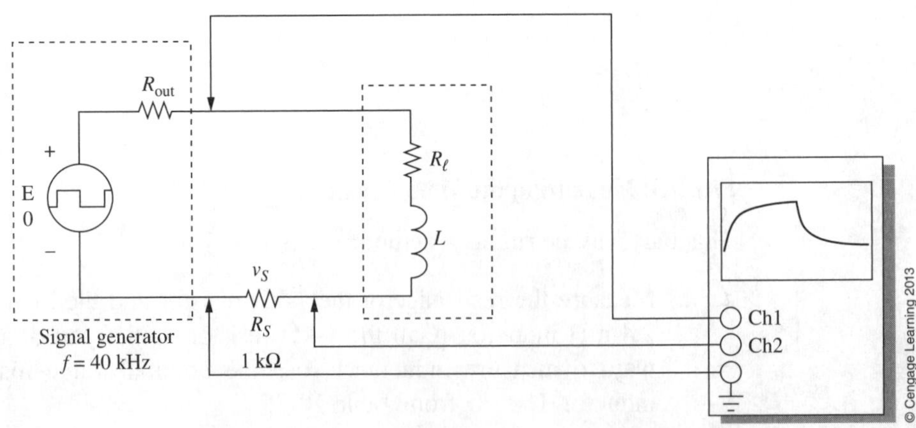

FIGURE 10-6 Current waveform for Test 5. The scope is triggered on Ch1 and the display is on Ch2. $R_T = R_{out} + R_\ell + R_S$.

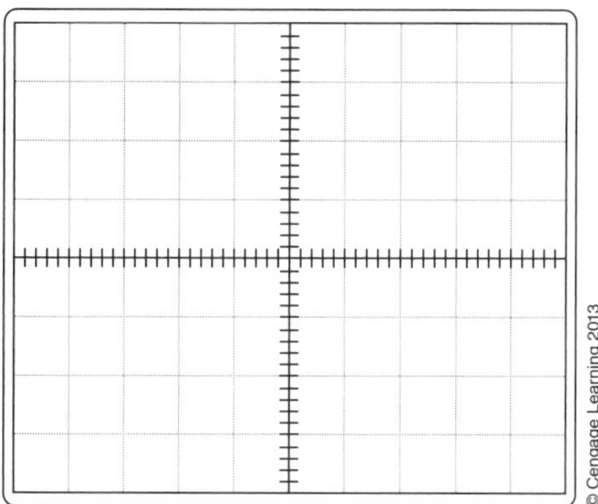

FIGURE 10-7 Waveform for Test 5.

COMPUTER ANALYSIS

6. Using either Multisim or PSpice, set up the circuit of Figure 10-6. To create a waveform equivalent to that used in Test 5, use a 40-kHz square wave with equal high and low times. If you are using PSpice, use pulse source VPULSE, whereas, if you are using Multisim, use the clock. (For reference, see Section 11.8 of the text.) To determine the amplitude for this pulse, note that in Test 5(c), the peak voltage across R_S is 5 V. This means that the peak current must be 5 mA. Use this to compute E. For example, if $R_{out} = 50\ \Omega$, coil resistance is 12 Ω, and $R_S = 1\ \text{k}\Omega$, then you need a pulse amplitude of $E = (5\ \text{mA})\ (50\ \Omega + 12\ \Omega + 1\ \text{k}\Omega) = 5.31\ \text{V}$.

   a. Run a transient analysis with the initial inductor current set to zero.

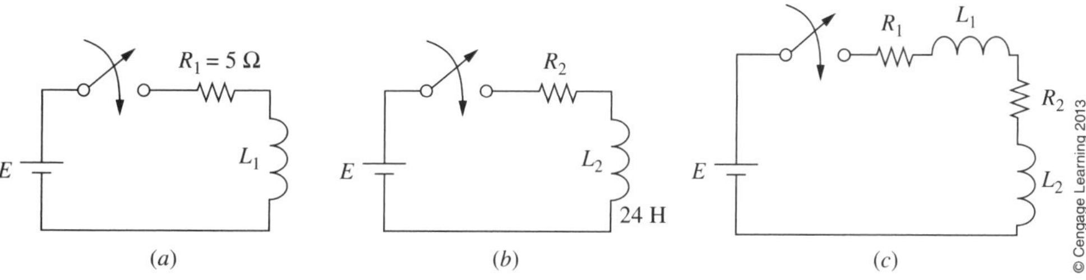

FIGURE 10-8

b. With the cursor, measure the time that it takes for current to rise to 63.2% of its final value. (This is τ.) Compare to the result of Test 5(e). Now measure the time constant on the decay portion and compare to Test 5(f).

PROBLEMS

7. For the circuit of Figure 10-2, if $E = 10$ V, $R_1 = 200\ \Omega$, $R_2 = 1200\ \Omega$, and $L = 5$ H,
   a. What voltage appears across the switch at the instant the switch is opened?
   b. How long will the transient last?
8. For the circuit of Figure 10-2, if $R_1 = 200\ \Omega$, $R_2 = 100\ \Omega$, and $L = 5$ H,
   a. What voltage appears across R_2 at the instant the switch is opened?
   b. How long will the transient last?
9. The circuit of Figure 10-8(a) takes 4 s to reach steady state, while that of (b) takes 12 s. The circuits of (a) and (b) are combined as in (c). How long will it take for the circuit of (c) to reach steady state?

FOR FURTHER INVESTIGATION AND DISCUSSION

Remember

Do not perform a full analytic transient analysis to determine the required information.

When you disturb an *RL* or an *RC* circuit you trigger a transient, creating voltages and currents that may greatly exceed the circuit's normal steady state values. Many times, however, it is only necessary to determine the general nature of the resulting waveforms and find values at key points—that is, you do not need a full transient solution. To help develop this idea, consider Figure 10-4. First, replace the 10-V source with a 250-mA dc current source, then install a switch in the leg containing R_4. Assume the coil resistance is negligible. Initially, the circuit is in steady state with the switch closed. You then open the switch. Prepare a written analysis of what happens using the following for guidelines:

a. Sketch the shape of the inductor current waveform from some time before the switch is operated to final steady state circuit operation. Using a series of diagrams, calculate the initial steady state current, the final steady state current, and the duration of the transient, and mark these on your diagram.
b. Sketch the coil voltage waveform.
c. (Optional) Calculate the peak value of the voltage spike and mark it on your diagram.

Name ______________________

Date ______________________

Class ______________________

The Oscilloscope (Part 1) Familiarization and Basic Measurements

OBJECTIVES

After completing this lab, you will be able to
- describe the operation and use of an oscilloscope,
- connect an oscilloscope to a circuit under test and select basic control settings,
- measure dc voltage,
- use an oscilloscope to observe time varying waveforms.

EQUIPMENT REQUIRED

- ☐ Oscilloscope
- ☐ Variable dc power supply
- ☐ Function generator
- ☐ Digital multimeter (DMM)

PRELIMINARY NOTE

Basic features of the oscilloscope are covered in *A Guide to Lab Equipment and Laboratory Measurements* at the beginning of this manual. You may wish to review this material before doing the lab.

PRE-STUDY (OPTIONAL)

Go to the Student Premium Website, choose *Lab Pre-Study—The Oscilloscope,* and download the material for Chapter 11.

EQUIPMENT USED

TABLE 11-1

Instrument	Manufacturer/Model No.	Serial No.
DMM		
Power Supply		
Function generator		
Oscilloscope		

DISCUSSION

> **Note**
>
> On traditional scopes, the "ground" is truly a ground in the sense that the scope chassis is grounded to earth. However, some modern scopes are isolated, and the "ground clip" is not really grounded. However, it is still the reference point that the probe tip measures voltage with respect to, and it is often simply referred to as a ground clip, even if it is not actually grounded. In this lab, we call it ground.
>
> In Lab 13, we look at alternate schemes, in particular, we study differential measurement schemes.

The oscilloscope is the key test and measurement instrument used for studying time varying waveforms. Its main feature is that it displays waveforms on a screen; with an oscilloscope, you can view and study waveforms, measure ac and dc voltages, frequency, period, phase displacement, and so on. However, the oscilloscope is a fairly complex instrument, and we therefore learn about it in stages. In this lab, we concentrate on operational procedures, front panel controls, and a few basic measurements; in Labs 12 and 13, we look at more advanced measurement techniques. Later labs add more detail.

Connecting to the Circuit under Test

Measuring a signal requires two connections, a probe tip connection and a reference connection (often simply referred to as a "ground" connection—see Note). The basic measurement connection is illustrated in Figure 11-1. The cable connecting the probe to the oscilloscope is usually a coaxial cable which helps shield the input from electrical noise pickup. The signal applied to the scope display electronics is thus the signal between the probe measurement tip and the ground reference clip.

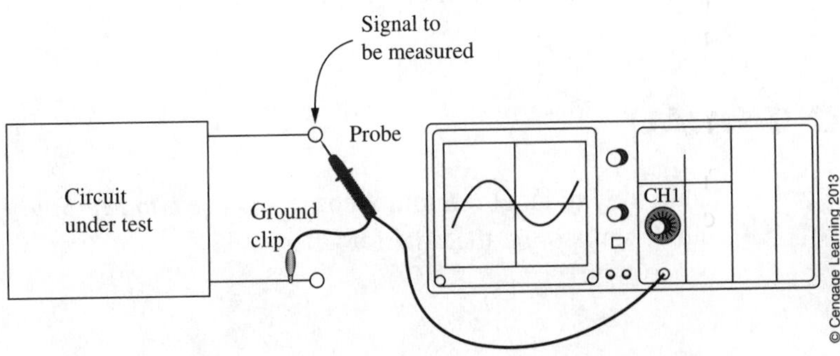

FIGURE 11-1 The basic oscilloscope measurement circuit.

Common probes are 1X and 10X (read as "times-one" and "times-ten"), but higher-attenuation probes, such as 20X, are also available. A 10X probe contains a 10:1 voltage divider, which attenuates the signal by a factor of 10; thus, when you use a 10X probe, you may have to multiply the scope readings by a factor of 10 to get the correct input (unless your scope automatically changes scales for you as most models now do).

Front Panel Controls

Front panel controls permit you to control the operation of your oscilloscope. Most oscilloscopes contain four main control sections—an input section, vertical and horizontal controls, and triggering controls.

Input Section and Controls

Most scopes are two- or four-channel instruments, with channels typically labeled *Ch1, Ch2,* and so forth. Each channel has buttons that permit you to select *AC* or *DC* coupling or *Ground* (sometimes designated *Gnd* or *0 V*). DC coupling displays the entire signal (dc plus any ac present), whereas ac blocks any dc component present and centers the resultant waveform about the 0 V axis. When set to *Gnd,* the input is grounded internally, thus permitting you to establish a *0-V* baseline (reference) on your screen. There may be other input control options, but these are the basic ones.

Vertical Controls

The main vertical control is the *Volts/Div* control. It permits you to set the Y-axis scale. For example, when set to 10V/Div, each major division on the Y-axis represents 10 volts. Each channel has its own independent controls. A secondary control, the Y-position control, is also provided. This control moves the entire waveform up or down to any position you want.

Horizontal Controls

The main horizontal control is the *Volts/Div* control. It permits you to set the times per division on the X-axis scale in s, ms, μs, or other units. For example, when set to 10ms/Div, each major division on the X-axis represents 10 ms. One control handles all channels. In addition, a horizontal position control permits you to move the waveform left or right to where you want it on the screen.

Trigger Controls

Triggering permits you to capture and display a stable waveform on your screen. Here are some of the controls that you need to know about:

Trigger Source: Selects the waveform from which to trigger, for example, *Ch1, Ch2, Ext* (external), and the AC line. When set to *Ch1,* for example, the waveform display is triggered from the input on *Ch1*. On high-end oscilloscopes, additional options are usually available.

Trigger Slope and Level: *Slope* selects whether the scope is to trigger on the positive or negative slope of the trigger source waveform, and *Level* selects the point on the waveform that you want triggering to occur. (Together, these define the basic trigger point definition, and most of the measurements that you do in this lab manual will use only this form of triggering.)

Trigger Mode: Modes include *auto, normal,* and *single sweep*. In the auto mode, the sweep always occurs, even with no trigger present; in the normal mode, a trigger must be present; in the single sweep mode (also called *single sequence*), a trigger is required but only one sweep results. (Other modes may be provided, but we will not consider them here.)

MEASUREMENTS

Caution

The ground points on oscilloscopes, power supplies, and other equipment are generally tied to the electrical power system ground through the U-ground pin on the electrical power outlet (plug-in). Since this connection ties all ground points together, you must be careful when connecting ground clips, as it is easy to inadvertently short out a component or even accidentally ground an output. While these grounds are required for safety reasons, they make poor signal paths. Therefore, be sure to use the ground clip supplied with the scope probe when making measurements.

PART A: Measuring dc Voltage with the Oscilloscope

We begin with measuring dc voltage with an oscilloscope. We will use both the visual (graticule) method and the cursor method here.

1. a. Connect a 1X probe to *Ch1* and set triggering to *Auto*. With 0 V applied, ensure the trace is on the Y-axis zero line. Set the scale to 1 V/Div.
   b. Connect the probe as in Figure 11-2 and set *VOLTS/DIV* to 1 V. With the voltmeter, set the power supply to 2 V and note the deflection on the screen. From this deflection, compute the measured voltage. (It should equal the applied voltage.) Record in Table 11-2.
   c. Now change *VOLTS/DIV* to 2 V, set $E = 4$ V, and note the position of the trace. Enter data in Table 11-2 and compute V. Repeat for $E = 15$ V at 5 *VOLTS/DIV*.
   d. Replace the probe with a 10X probe. Using the oscilloscope, set the supply successively to 10 V, 15 V, and 22.5 V. Record data, including the *VOLTS/DIV* settings that you choose. Compare to the meter reading.
   e. If your scope has a cursor, repeat the preceding measurement using the cursor.

TABLE 11-2

Probe	Input Voltage	VOLTS/DIV Setting	Voltage from Graticule Method	Voltage from Cursor Method
1X	2 V	1-V		
1X	4 V	2-V		
1X	15 V	5-V		
10X				10 V
10X				15 V
10X				22.5 V

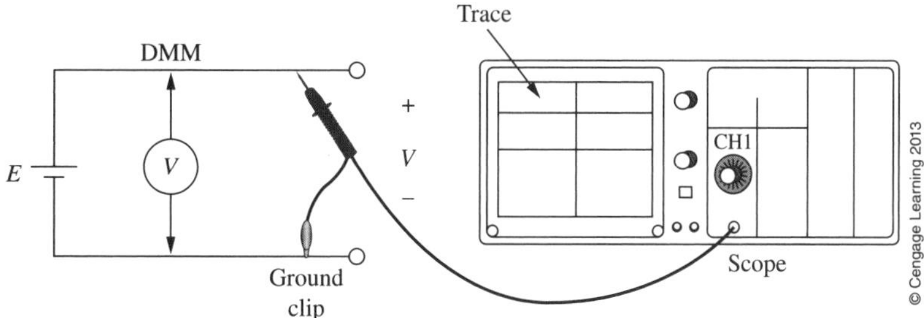

FIGURE 11-2 Circuit for Test 1.

2. Move the input coupling switch to *ac*. Set the supply to the voltages of Table 11-2. What happens? Why?

____

> **Note**
>
> If your power supply output cannot be floated, you may not be able to do this test.

3. Return coupling to *dc*. Now set the trigger to *normal*. Note that the trace disappears. (This is because there is no trigger point in dc to start the sweep.) Return the trigger to *auto*.

4. Make sure your power supply output is floating—see Note. Reverse the probe connections to the power supply. Adjust the power supply voltage up and down. Note the deflection on the screen. Describe what happened.

____

PART B: Observing Waveforms

Preliminary: When observing ac waveforms, you need to make sure the scope probe is properly compensated. To do this, connect the probe to the channel that you intend to use, then touch the probe tip to the *calibration test point* (sometimes called *Probe Check*) on the front panel of your oscilloscope. Adjust the *VOLTS/DIV* control, the *SEC/DIV* control, and the *trigger controls* until you get the calibration waveform on your screen. (It should be a square wave.) If necessary, adjust the probe to get a proper square wave. (On some probes, you twist the probe barrel; on others, you adjust a compensating screw.)

The Test Setup

Replace the dc source of Figure 11-2 with a signal generator. Set your scope to *ac* coupling, triggering to positive slope, and mode to *Normal*.

5. a. Set the generator to a 2-kHz sine wave with 2 V amplitude. Set the scope *Timebase* to 100 μs/Div and the vertical control to 1 V/Div. Sketch the waveform in Figure 11-3(a).
   b. Change the trigger slope to negative and repeat (a). Sketch the waveform as Figure 11-3(b).
   c. Return the trigger slope to positive, then set input coupling to *dc*. (Make sure the dc offset on your function generator is set to zero.) What happens to the waveforms? (Compare to those of Figure 11-3.)

____

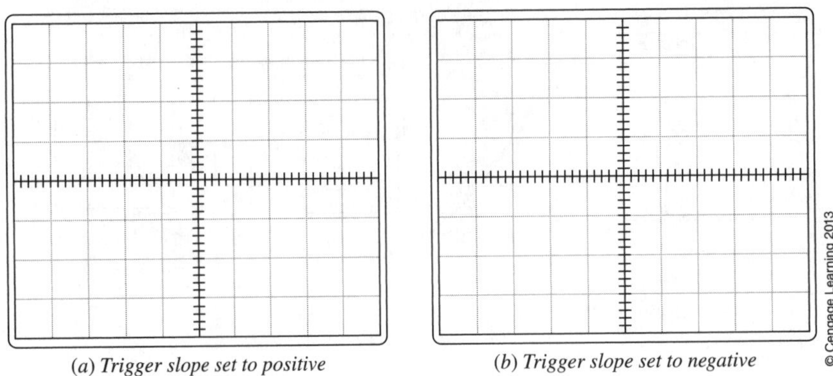

(*a*) *Trigger slope set to positive* (*b*) *Trigger slope set to negative*

FIGURE 11-3 One cycle of a sine wave.

d. Return the coupling to *ac*. Set f = 500 Hz and change the time base to 500 μs/Div. Sketch here.

6. Repeat Steps 5(a) and (b) for a square wave and for a triangular wave. Sketch waveforms as Figures 11-4 and 11-5.

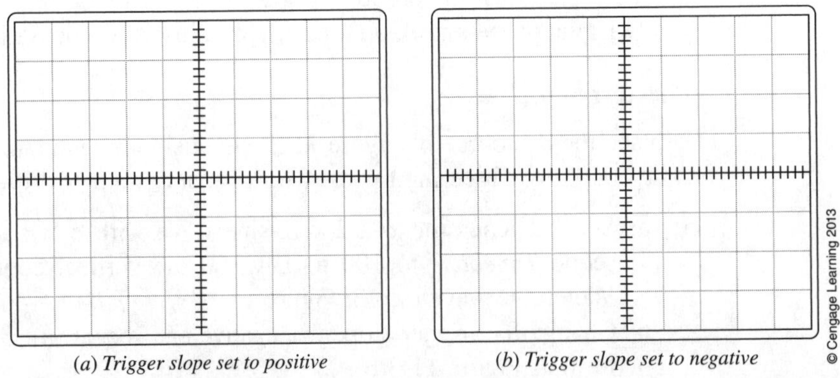

(*a*) *Trigger slope set to positive* (*b*) *Trigger slope set to negative*

FIGURE 11-4 One cycle of a square wave.

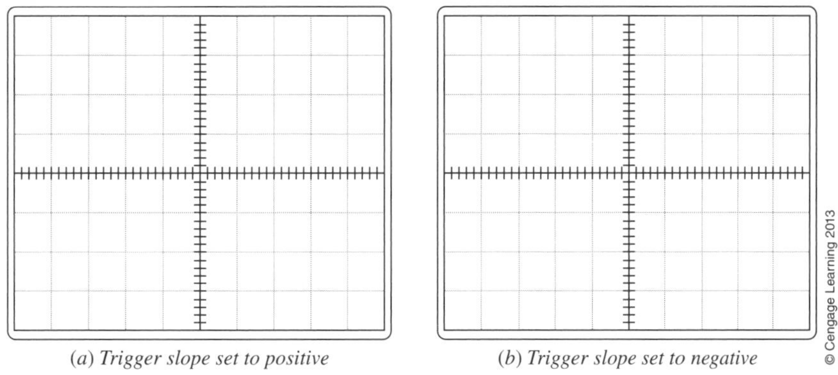

FIGURE 11-5 One cycle of a triangular wave.

PROBLEMS

7. With a 1X probe and *VOLTS/DIV* (vertical sensitivity) set at 5 V/Div, a dc voltage moves the trace up 3.4 grid lines. What is the input voltage? __________

8. With a 1X probe and the vertical sensitivity set at 10 V/Div, a dc voltage moves the trace down by 1.2 grid lines. What is the input voltage? __________

9. With a 10X probe and the vertical sensitivity control set at 5 V/Div, a dc voltage moves the trace up 2.5 grid lines. What is the input voltage? (There are two possible answers, depending on your oscilloscope—see earlier Note. Answer in terms of the scope you used in this lab.)

Name ______________________

Date ______________________

Class ______________________

Basic ac Measurements: Period, Frequency, and Voltage (The Oscilloscope—Part 2)

OBJECTIVES

After completing this lab, you will be able to use an oscilloscope to

- measure period and frequency of an ac waveform,
- measure amplitude and peak-to-peak voltage,
- measure instantaneous voltage,
- determine the equation for a sinusoidal voltage from the oscilloscope readings.

EQUIPMENT REQUIRED

- ☐ Oscilloscope
- ☐ Signal or function generator

PRE-STUDY (OPTIONAL)

Go to the Student Premium Website, choose *Lab Pre-Study—The Oscilloscope,* and download the material for Chapter 12.

EQUIPMENT USED

TABLE 12-1

Instrument	Manufacturer/Model No.	Serial No.
Oscilloscope		
Signal or function generator		

© Cengage Learning 2013

TEXT REFERENCES

Section 15.4 FREQUENCY, PERIOD, AMPLITUDE, AND PEAK VALUE
Section 15.5 ANGULAR AND GRAPHIC RELATIONSHIPS FOR SINE WAVES
Section 15.6 VOLTAGES AND CURRENTS AS FUNCTIONS OF TIME

DISCUSSION

Digital oscilloscopes do many things automatically—for example, they measure frequency and period without any intervention from the user. However, first and foremost, an oscilloscope is a visualization instrument, and we feel that visualization helps you gain a better understanding of what the scope is all about. Accordingly, as our starting point, we ask you to make some measurements directly from your screen using its graticule system. However, as you gain familiarity and put the oscilloscope to work in later labs, we shift more of these tasks over to cursors and other more accurate methods.

Measuring Frequency and Period: The frequency of a waveform is its number of cycles per second, and its period is the duration of one cycle. On digital scopes, these parameters can usually be measured automatically by the scope itself. However, you can also use the X-axis time scale to measure the period directly, then use the relationship $f = 1/T$ to compute frequency. For example, if the time base is set to 20 μs per division and one cycle is 4 divisions, then $T = 4(20\ \mu s) = 80\ \mu s$ and $f = 1/80\ \mu s = 12.5$ kHz.

Measuring Voltage: An oscilloscope displays the instantaneous value of its input voltage. Thus, an oscilloscope may be used to measure peak voltage, peak-to-peak voltage, and indeed, the voltage at any point on a waveform. A convenient way to do this is to use the scope's cursor. If you do not have a cursor, visually determine the deflection of the trace at that point, then multiply by the vertical sensitivity setting.

Equations for Sinusoidal Voltage from Oscilloscope Readings: Mathematically, the voltage at any point on a sine wave can be found from the equation

$$v = V_m \sin \alpha \qquad (12\text{-}1)$$

where α is the angular position on the cycle as indicated in Figure 12-1. If you know V_m, you can determine the voltage at any position by direct substitution into Equation 12-1. (Since one cycle represents 360°, one half-cycle represents 180°, one-quarter cycle represents 90°, and so on. The angular position at any other point can be determined by direct proportion.)

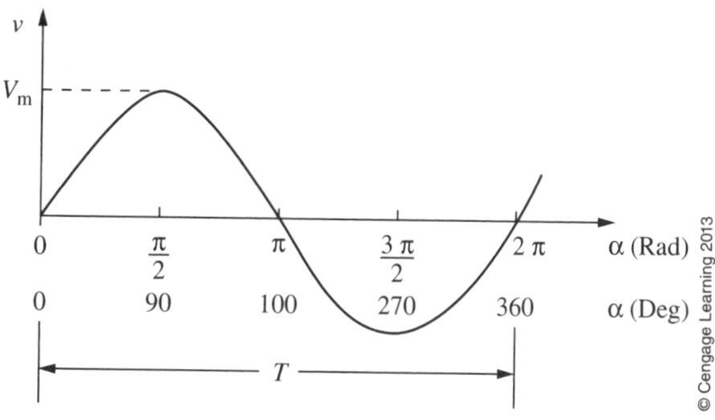

FIGURE 12-1 A sinusoidal waveform.

Voltage as a Function of Time: Equation 12-1 may be rewritten as a function of time as

$$v(t) = V_m \sin\omega t \tag{12-2}$$

where $\omega = 2\pi f = 2\pi/T$ and t is time measured in seconds. By measuring V_m and T on the screen, you can use Equation 12-2 to establish the analytic expression for any measured sinusoidal voltage.

Control Settings: Always select *VOLTS/DIV* and *SEC/DIV* settings to yield best results. For example, if you set the amplitude and time scales to spread a waveform over the screen, you can get more accurate measurements than if you compress the waveform into a tiny space. (Most settings here and in future labs will be left for you to select.)

MEASUREMENTS

PART A: Measuring Period and Frequency

1. a. Connect the oscilloscope to the signal generator. Set the oscilloscope time base to 100 μs/Div, coupling to *ac*, and the trigger slope to *positive*. Set the signal generator to obtain a sine wave that is five divisions in length. Thus, T = ___________ and f = ___________. Compare f to the frequency set on the generator dial.

   ____

   b. Repeat for a time base setting of 20 μs per division and two cycles in eight divisions. T = ___________ and f = ___________. Compare to the frequency set on the signal generator dial.

   ____

PART B: Amplitude and Peak-to-Peak for a Sine Wave

2. For best results, set peak-to-peak amplitude rather than zero to peak. Use the vertical control to position the waveform between grid lines if necessary. For this test, set the vertical sensitivity to 0.5 V/Div and choose a frequency of 2 kHz.

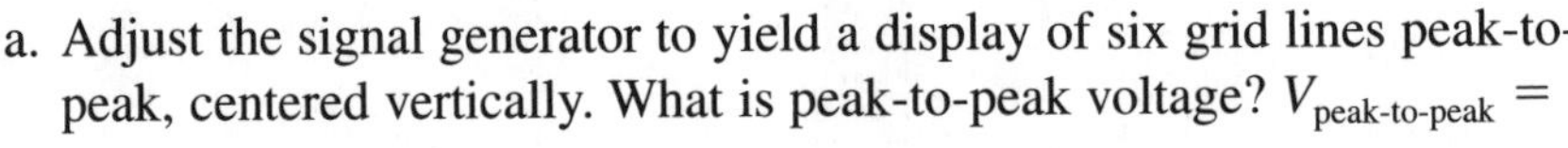

a. Adjust the signal generator to yield a display of six grid lines peak-to-peak, centered vertically. What is peak-to-peak voltage? $V_{\text{peak-to-peak}}$ = ______________.

b. What is V_m? V_m = ______________.

c. Sketch the waveform below with V_m and peak-to-peak voltages carefully labeled.

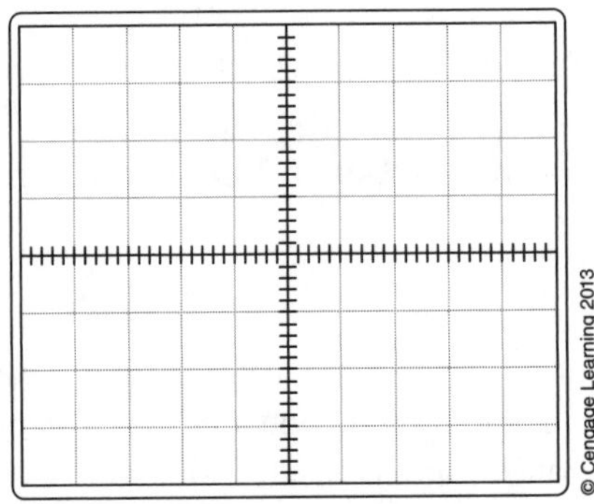

PART C: Instantaneous Value of a Sine Wave

3. a. Set the vertical sensitivity control to 1 V/Div, the time base to 100 μs/Div, and obtain a waveform on the screen with a cycle length of exactly eight divisions. Adjust the signal generator for a peak-to-peak display of four grid lines. Sketch the waveform as Figure 12-2. Label the vertical and horizontal axes in volts and ms.
   b. Using your scope's cursor, measure the voltage at 100-μs intervals and record in Table 12-2.

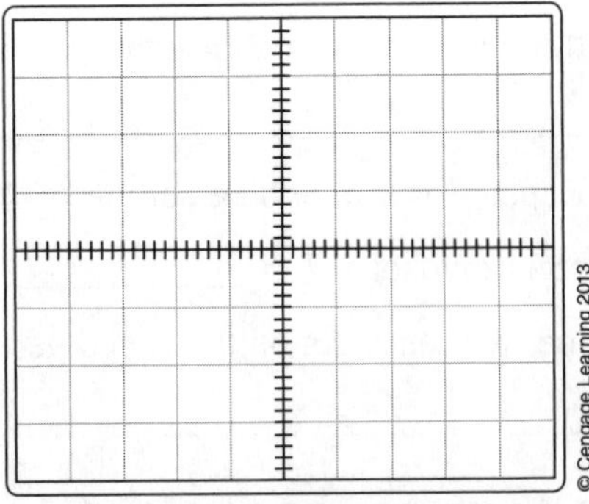

FIGURE 12-2 Measured waveform for Test 3.

c. How many degrees does each 100-μs division represent?

_______________.

Record the value of α for each value of t shown in Table 12-2.

d. Using Equation 12-1, verify each entry in the table. (Show a few sample calculations.)

TABLE 12-2

		Voltage	
t (ms)	α (deg)	Measured	Computed
0			
0.1			
0.2			
0.3			
0.4			
0.5			
0.6			
0.7			
0.8			

PART D: Equation for Sinusoidal Voltage as a Function of Time

4. a. For the waveform of Figure 12-2, determine ω.

ω = _______________

b. Using the measured values of ω and V_m, write the equation for the voltage in the form of Equation 12-2.

$v(t)$ = _______________

c. Using this equation, compute v at t = 50 μs and t = 150 μs. Show details below.

PROBLEMS

5. With the time base set to 500 ns/div, four cycles of a waveform occupies 10 divisions. What is the period and the frequency of the waveform?

   Period ____________ Frequency ____________

   Amplitude ____________

6. Given $v(t) = 100 \sin 377t$

   a. What is the value of V_m? $V_m =$ ____________

   b. What are the frequency and period?

      $f =$ ____________ $T =$ ____________

   c. Compute the voltage at $t = 20$ ms. Sketch the waveform and show where $t = 20$ ms occurs on the waveform.

7. Determine the equation $v(t)$ for the voltage depicted in Figure 12-3.

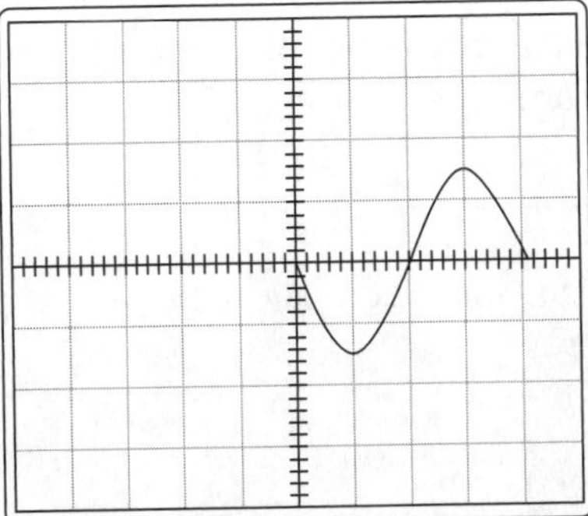

Vertical = 5V/Div
Time base = 0.1 µs/Div

FIGURE 12-3 Assume the point where the waveform starts is $t = 0$ s.

Name ____________________

Date ____________________

Class ____________________

LAB 13

ac Voltage and Current (The Oscilloscope—Part 3: Additional Measurement Techniques)

OBJECTIVES

After completing this lab, you will be able to use an oscilloscope to

- measure rms values for sinusoidal voltage,
- measure superimposed ac and dc voltages,
- measure ac current using a sensing resistor,
- display two waveforms simultaneously on a dual channel oscilloscope,
- measure phase displacement with a dual channel oscilloscope,
- measure voltage using differential measurement techniques.

EQUIPMENT REQUIRED

- ☐ Dual channel oscilloscope
- ☐ Signal or function generator
- ☐ Digital multimeter (DMM) and VOM (optional)

COMPONENTS

☐ Resistors:	10-Ω, 82-Ω, 100-Ω, 150-Ω, 1-kΩ, 2-kΩ, 3.3-kΩ, 5.6-kΩ, 6.8-kΩ, 10-kΩ (all 1/4-W)
☐ Capacitor:	0.01-μF
☐ Battery:	1.5-V

PRE-STUDY (OPTIONAL)

For Multisim-based pre-study simulations, go to the Student Premium Website, choose *Lab Pre-Study—The Oscilloscope,* and download the simulations for Chapter 13.

EQUIPMENT USED

TABLE 13-1

Instrument	Manufacturer/Model	Serial No.
DMM		
VOM (optional)		
Dual channel oscilloscope		
Signal of function generator		

TEXT REFERENCE

Section 15.6 VOLTAGES AND CURRENTS AS FUNCTIONS OF TIME
Section 15.9 EFFECTIVE (RMS) VALUES

DISCUSSION

Measuring the rms Voltage of a Sine Wave. An oscilloscope may be used to determine the rms value of a sinusoidal voltage since, for a sine wave, $V_{rms} = 0.707\ V_m$ and V_m can be measured directly on the screen. One of the reasons you might do this is convenience—if you have a signal already displayed on the screen, there is no need to connect a voltmeter. However, a more fundamental reason has to do with frequency—most meters have very limited frequency ranges. For example, many DMMs can measure to only a few kHz (although some can measure to several hundred kHz), while analog VOMs typically can measure up to about 100 kHz. (Check your meter's specs.) On the other hand, even inexpensive oscilloscopes can measure to the tens of MHz, while top-of-the-line units can measure to hundreds of MHz, or even to the GHz range.

Measuring Current. Oscilloscopes can also be used to measure current, although not directly. There are, however, two indirect ways to measure current. The first is to insert a small, known resistor (sometimes called a *sensing resistor*) into the circuit, measure the voltage across it using an oscilloscope, then use Ohm's law to compute current. (This method is inexpensive and widely used. The other method uses a *current probe,* but few introductory courses have access to equipment of this type so we will not consider it.)

The current-sensing resistor approach is based on Ohm's law. For a purely resistive circuit, $v = Ri$. Thus, if voltage is sinusoidal, current will be sinusoidal also and vice versa—that is, v and i are in phase. Therefore, if voltage v is

$$v = V_m \sin \omega t$$

then

$$i = I_m \sin \omega t$$

Current can thus be determined by measuring V_m and computing I_m from $I_m = V_m/R$. If the rms value of current is needed, it may be determined from the equation $I_{rms} = 0.707\ I_m$.

Dual Channel Measurements. With a dual channel oscilloscope, you can display two waveforms simultaneously. This permits you to determine phase relationships between signals, compare wave shapes, and so on. (Many modern oscilloscopes have more than two channels, in which case they are more properly called multichannel scopes.)

Differential Voltage Measurements. Sometimes you need to measure the voltage across a component where the normal technique of placing the probe tip at one end and the ground clip at the other end shorts out part of the circuit. For problems such as this, *differential measurement* can be used. In Part E of this lab, you learn how make such measurements.

MEASUREMENTS

PART A: RMS Values and the Frequency Response of ac Meters

We begin with a look at the frequency response of various instruments. Here, we compare the ability of the oscilloscope, DMM, and VOM (if you have one) to measure voltage at different frequencies.

1. a. Assemble the circuit of Figure 13-1. Adjust the signal generator to a 100 Hz sine wave with 12 V peak-to-peak (i.e., $V_m = 6$ V. Thus, $V_{rms} = 0.707 \times 6 = 4.24$ V. This is recorded in Table 13-2 as *Actual rms*.) Now measure and record the rms voltage using the DMM and the VOM.
   b. Repeat step (a) at the other frequencies indicated in Table 13-2.

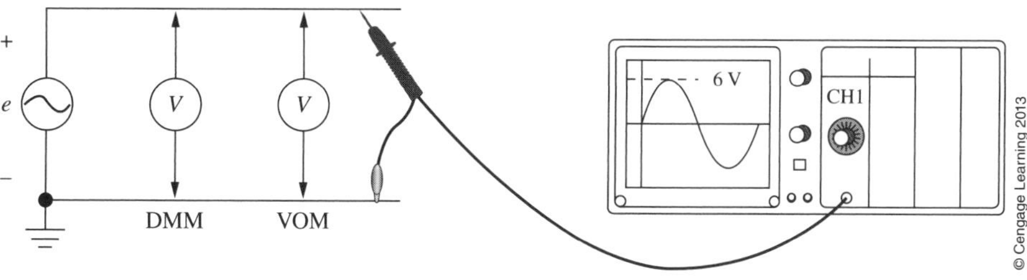

FIGURE 13-1 Circuit for Test 1.

TABLE 13-2

Frequency	Scope Reading	Actual rms	DMM Reading	VOM Reading
100 Hz	6 V	4.24 V		
1000 Hz	6 V	4.24 V		
10,000 Hz	6 V	4.24 V		
100 kHz	6 V	4.24 V		
1 MHz	6 V	4.24 V		

c. What conclusion do you draw from the data of Table 13-2?

____

____

PART B: Superimposed ac and dc

2. Add the 1.5-V battery to the circuit as in Figure 13-2(a). Set the signal generator to a 100-Hz sine wave. Select *ac* coupling (to temporarily block the dc component while you set the ac component) and adjust the output of the generator to $V_m = 2$ V (i.e., 4 $V_{p\text{-}p}$). Return coupling to *dc*.
   a. You now have superimposed ac and dc voltages. Sketch as Figure 13-2(b).
   b. Compute the true rms value for this waveform from $V = \sqrt{V_{dc}^2 + V_{ac}^2}$ where V_{dc} is the dc component of the waveform and V_{ac} is the rms value of its ac component.

   $V =$ __________

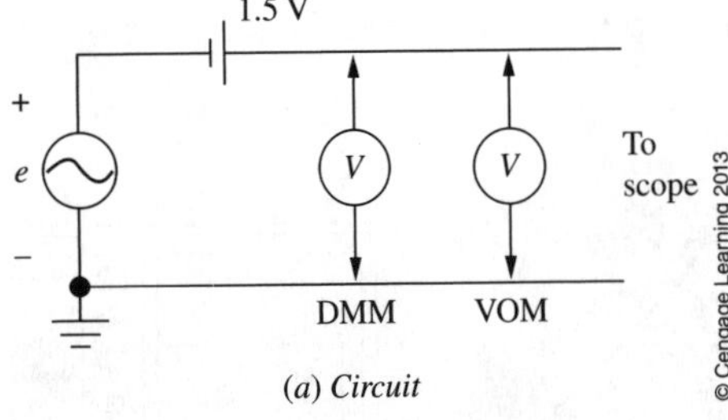

(*a*) Circuit (*b*) Waveform

FIGURE 13-2 Circuit for Test 2.

c. Measure the voltage using the meters. (If you have a true rms reading DMM, use it to measure the voltage as well.)

DMM reading ____________ VOM reading ____________

d. Discuss the results of (b) and (c). Note: If you have a true rms reading DMM, its result should agree with your calculation. However, your average responding meter's result will not.

____

____

____

PART C: Dual Channel Measurements

3. Assemble the circuit of Figure 13-3. Set the signal generator to a 2.5-kHz sine wave with $V_m = 2$ V. Set your scope to 50 μs/Div and view the waveforms.
   a. Assuming *Ch1* as the reference voltage, it can be shown that the voltage of *Ch2* will lag *Ch1* by

   $$\theta = \left[\tan^{-1}\frac{1}{\omega RC}\right] - 90° =$$

   Using this equation, compute angle θ.
   b. Reset the time base scale to 20 μs/Div to expand the trace. Measure the displacement between the zero crossover points using cursors, and record as Δt.

   $\Delta t =$ ____________

   c. Using the value of Δt and the period of the waveforms, compute the measured angle θ.

   $\theta =$ ____________

   d. How does the measured value for θ compare to the computed value?

   ____

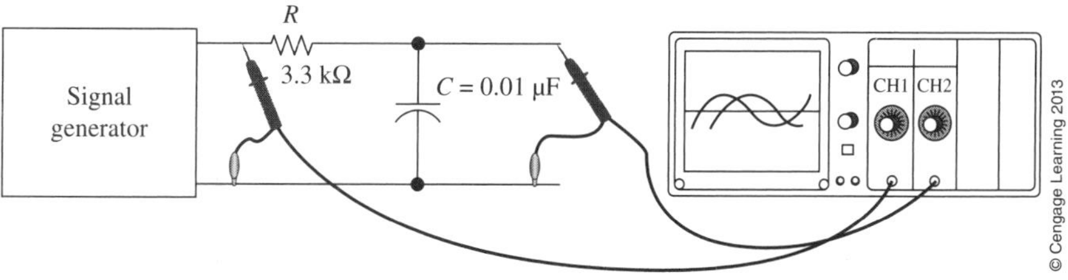

FIGURE 13-3 Circuit for Test 3. Use $R = 3.3$ kΩ and $C = 0.01$ μF.

e. Taking Ch1 as reference, sketch the waveforms below.

f. Write the equations for the two voltages using the measured values.

$v_1(t) =$ __________

$v_2(t) =$ __________

PART D: Current Measurement with an Oscilloscope

Consider Figure 13-4(a). The load current I_L is given by $I_L = E/R_L$ where E and I_L are the rms values of the source voltage and load current, respectively. To measure this current using an oscilloscope, you can add a sensing resistor R_S as in Figure 13-4(b), then measure the voltage across it, convert to rms, then compute current as $I_L' = V_S/R_S$. (As long as R_S is small compared to R_L, the accuracy will be good.)

4. a. Accurately measure the 10-Ω sensing resistor and R_L. Set up the circuit of Figure 13-4(b). (Note the placement of the sensing resistor. With the resistor placed as shown, you can connect both ground clips of the oscilloscope directly to the ground of the signal generator without fear of ground problems.) Use a 100-Hz sinusoidal source. Using Ch1, set E_m to 3 V (i.e., 6 $V_{p\text{-}p}$).
   b. Measure the voltage across the sensing resistor using Ch2, then convert to rms.

   $V_S =$ __________ (rms)

   c. Using Ohm's law, determine the rms value of the measured load current

   $I_L' = V_S/R_S =$ __________ (mA rms)

   Verify this current by comparing it to that measured by the meter.
   d. The load current that you are trying to measure is $I_L = E/R_L$, where E is the rms value of the source voltage. Using the measured value of R_L,

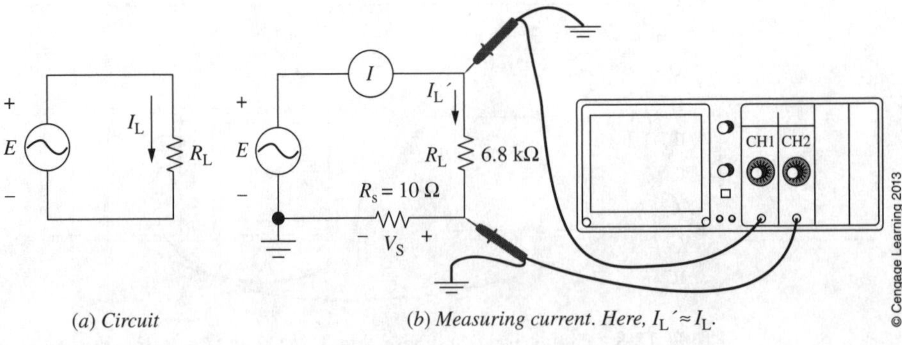

(a) Circuit

(b) Measuring current. Here, $I_L' \approx I_L$.

FIGURE 13-4 Circuit for Test 4. Use $R_S = 10\ \Omega$.

compute I_L using this formula and compare to the current determined by the sensing resistor approach in (c).

5. Replace R_L of Figure 13-4 with the network of Figure 13-5 and repeat steps 4(b) to 4(d).

V_S = __________ (rms voltage across R_S); $I_L' = V_S/R_S$ = __________

I_L = __________ (Determined by circuit analysis)

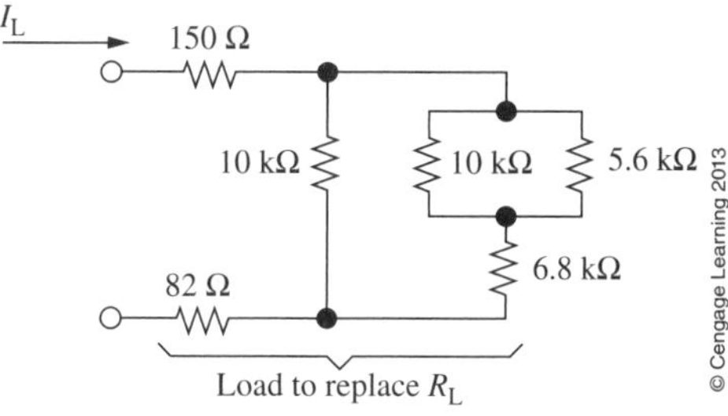

FIGURE 13-5 Circuit for Test 5. Replace R_L with this network.

How do the results compare?

PART E: Differential Measurements Using a Traditional, Grounded Oscilloscope

> **Notes**
>
> With some modern oscilloscopes, there are easier approaches. For example, floating, battery-operated portable scopes permit the measurement scheme of Figure 13-6.

Assume that the oscilloscope of Figure 13-6 is a traditional, grounded chassis scope. Suppose you want to measure the voltage across resistor R_1 using such an oscilloscope. As a first thought, you might try the connection shown. With this connection, however, the ground lead shorts out R_2. Alternatively, you might try reversing the probe tip and ground clip connections. However, this shorts the source output to ground. Thus, neither approach is usable. However, there is a simple solution—you can use differential measurement as illustrated in Figure 13-7. (See also margin Note.)

In Figure 13-7, Ch1 measures the voltage from point *a* to ground while Ch2 measures the voltage from point *b* to ground. The oscilloscope has mathematical capability (or on older scopes, a differential mode) that permits you to display Ch1 minus Ch2. This yields a display of V_{ab}, the voltage between points *a* and *b*. Let us investigate.

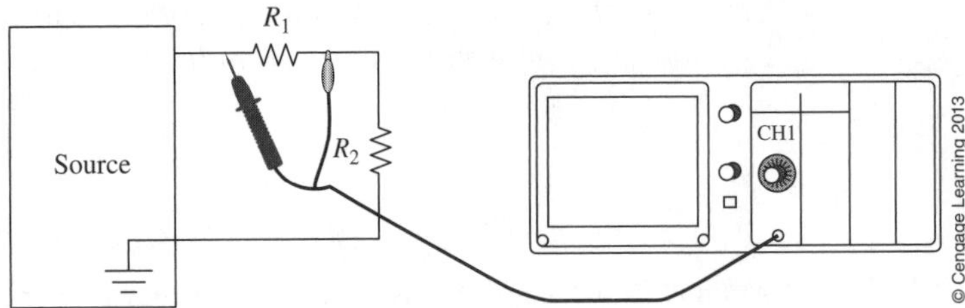

FIGURE 13-6 An incorrect way to view the voltage across R_1 using a traditional oscilloscope.

6. a. Assemble the circuit of Figure 13-7. Set the signal generator to a 1-kHz sine wave.
   b. Set the source voltage to V_m to 3 V and select an appropriate scale for easy viewing. Use the *Ch1* cursor to confirm V_m.
   c. Set *Ch2 Volts/Div* to the same setting used for *Ch1*. Use the *Math* function on your scope to select and display *Ch1–Ch2* (or if you have an older scope, select its *difference* mode). This is voltage V_{ab}.
   d. Using the voltage divider rule, compute V_{ab}, and compare to the value measured in step (c).
7. We will now look at what happens if you had tried to use the circuit of Figure 13-6 to measure the voltage across R_1.
   a. Consider again the circuit of Figure 13-7. Disconnect Probe 2 (i.e., Ch2) and set the channel select back to Ch1. You should see 3 V (i.e., 6 $V_{p\text{-}p}$) on the screen since Ch1 is still measuring full source voltage.
   b. Move the ground clip of Probe 1 to point *b* and note the scope display.

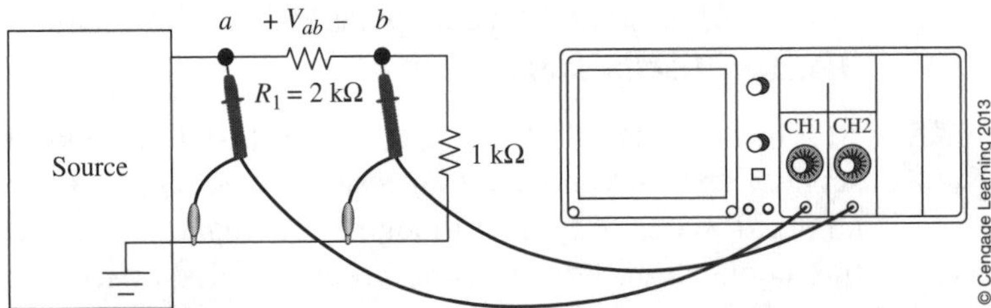

FIGURE 13-7 The correct way to view the voltage using differential measurement.

c. In Test 7(b), your probe is connected from *a* to *b;* thus, you are measuring V_{ab}. However, it is different than V_{ab} you measured in Test 6(d). Briefly discuss what you are now seeing.

PROBLEMS

8. Given: $v_1(t) = 100 \sin \omega t$ and $v_2(t) = 80 \sin(\omega t + 30°)$ displayed on a scope screen with v_1 as reference. If $f = 100$ Hz, sketch the oscilloscope trace in the space below as Figure 13-8(a).
9. Repeat Question 8 [as Figure 13-8(b)] if v_2 is the reference waveform.

FIGURE 13-8 (*a*) *Waveform for Question 8* (*b*) *Waveform for Question 9*

Name ______________________

Date ______________________

Class ______________________

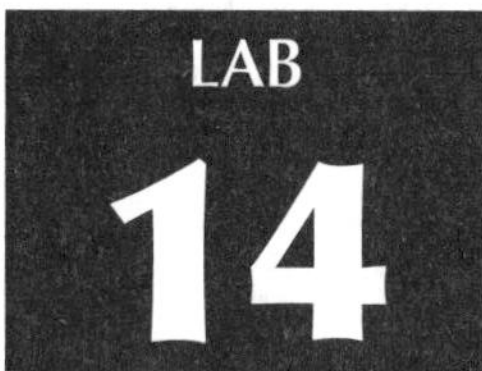

Capacitive Reactance

OBJECTIVES

After completing this lab, you will be able to

- measure phase difference between voltage and current in a capacitance,
- measure capacitive reactance and verify theoretically,
- determine the effect of frequency on capacitive reactance.

EQUIPMENT REQUIRED

- ☐ Dual channel oscilloscope
- ☐ Signal or function generator
- ☐ Digital multimeter (DMM) (two)

COMPONENTS

- ☐ Resistors: 10-Ω, 220-Ω, 1/4-W
- ☐ Capacitor: 1.0-μF (two), nonelectrolytic

PRE-STUDY (OPTIONAL)

Interactive simulation *CircuitSim 16-4* may be used as pre-study for this lab. From the Student Premium Website, select *Learning Circuit Theory Interactively on Your Computer,* download and unzip *Ch 16* (if you have not already done so), and run the corresponding *Sim* file. (Note that its circuits are similar to but not identical to those of this lab.)

EQUIPMENT USED

TABLE 14-1

Instrument	Manufacturer/Model No.	Serial No.
DMM #1		
DMM #2		
Dual channel oscilloscope		
Signal or function generator		

TEXT REFERENCE

Section 16.6 CAPACITANCE AND SINUSOIDAL ac

DISCUSSION

Capacitance and Sinusoidal ac: When an ideal capacitance is connected to a sinusoidal voltage source, current leads by 90° as illustrated in Figure 14-1. Thus, if

$$v_C = V_m \sin \omega t \qquad (14\text{-}1)$$

then

$$i_C = I_m (\sin \omega t + 90°) \qquad (14\text{-}2)$$

where

$$I_m = V_m / X_C \qquad (14\text{-}3)$$

The quantity X_C is termed *capacitive reactance* and is given by the formula

$$X_C = 1/\omega C \ \Omega \qquad (14\text{-}4)$$

where $\omega = 2\pi f$ rad/s.

Practical Note

V_C and I_C are the values that you read on meters, while V_m and I_m are the values that you read on the oscilloscope.

Effective (rms) Voltage and Current Relationships: In practice, we usually use effective values rather than peak values. Their ratios are the same, however. Thus, reactance can also be expressed as

$$X_C = V_C / I_C \ \Omega \qquad (14\text{-}5)$$

where V_C and I_C are the rms values of v_C and i_C, respectively.

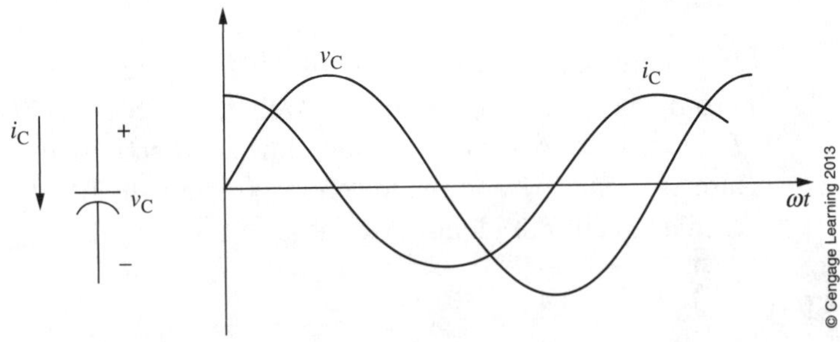

FIGURE 14-1 Voltage and current for capacitance.

MEASUREMENTS

PART A: Phase Relationships for Capacitance

Consider Figure 14-2. Voltage v_S across R_S is in phase with capacitor current i_C. Provided circuit resistance is small compared to X_C, source voltage is approximately equal to the v_C. Thus, the phase difference between Ch1 and

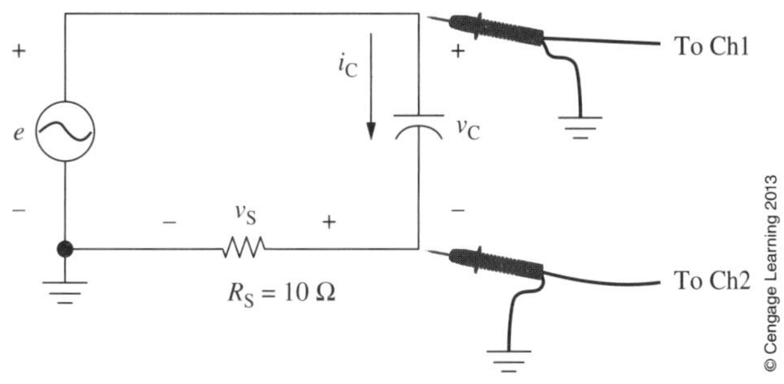

FIGURE 14-2 Circuit resistance is very small. Therefore, $v_C \approx e$.

Ch2 is approximately equal to the angle between v_C and i_C. It should be close to 90°. Current should lead.

1. a. Assemble the circuit using $R_S = 10\ \Omega$ and $C = 1.0\ \mu F$.
   b. Set the signal generator to a 500-Hz sine wave. Sketch the waveforms below, appropriately labeled as v_C and i_C. Determine the phase shift between v_C and i_C and indicate on the diagram. How close is it to the theoretical value of 90°?

   c. Vary the frequency a few hundred Hz up and down and observe the phase shift. Is there any appreciable change? Based on this observation, state in your own words the relationship between voltage across and current through capacitance.

____

____

PART B: Magnitude Relationships for Capacitance

2. a. Replace R_S of Figure 14-2 with an accurately measured 220-Ω resistor.

   R_S = ___________

Practical Notes

1. For this lab, you need a DMM with a frequency response up to 1 kHz. (Most DMMs can handle this range.)
2. Since reactance depends on frequency, accurate frequency settings are essential. However, the frequency dial on some signal generators is not accurately calibrated, and you may require the oscilloscope to set the frequency. (If you have a generator with a digital readout, the scope should not be necessary.)

b. Set the signal generator to 500 Hz at $V_m = 6$ V. With a meter (or an oscilloscope if necessary), carefully measure the voltage V_C and the voltage V_S. Calculate current I_C using Ohm's law and the measured value of V_S. Record as rms values.

$V_C =$ ______________ $V_S =$ ______________ $I_C =$ ______________

c. Calculate the reactance of the capacitor using the measured V_C and I_C.

$X_{C(\text{measured})} = V_C/I_C =$ ______________

d. If possible, measure C using a bridge. (Otherwise, use the nominal value.) Compute X_C using Equation 14-4. Compare to the measured value of 2(c).

$C =$ ______________ $X_{C(\text{computed})} =$ ______________

3. a. Add the second capacitor in series and determine the equivalent reactance of the series combination using the same procedure that you used in Test 2(b) and (c).

$V_C =$ ______________ $V_S =$ ______________

$I_C =$ ______________ $X_{\text{eq(measured)}} =$ ______________

b. If possible, measure the second capacitor and compute C_{eq} for the series combination. $C_{eq} =$ ______________. Use Equation 14-4 to determine $X_{\text{eq(computed)}}$. $X_{\text{eq(computed)}} =$ ______________. Compare to the measured value of 3(a).

4. a. Connect both capacitors in parallel and measure total reactance X_T using the same procedure that you used in Test 2(b) and (c).

$V_C =$ ______________ $V_S =$ ______________

$I_C =$ ______________ $X_{T(\text{measured})} =$ ______________

b. Compute C_T for the parallel combination. C_T = ______________.

Now use Equation 14-4 to determine $X_{T(computed)}$. $X_{T(computed)}$ = _________.

Compare to the measured value of 4(a).

PART C: Variation of Reactance with Frequency

5. a. Replace the parallel combination with a single 1.0-μF capacitor and measure V_C and V_S and compute reactance (using the procedure of Test 2) at each of the frequencies listed in Table 14-2. Record as $X_{C(measured)}$.

   b. Using Equation 14-4, compute reactance at each of the frequencies and record as $X_{C(computed)}$.

TABLE 14-2

f (Hz)	V_C	V_S	I_C	$X_{C(measured)}$	$X_{C(computed)}$
100					
200					
300					
400					
500					
600					
700					
800					
900					
1000					

   c. Using the graph sheet of Figure 14-3, plot measured and computed reactances versus frequency. Label each plot.

   d. Analyze the results. That is, comment on how well the measured values agree with the computed values.

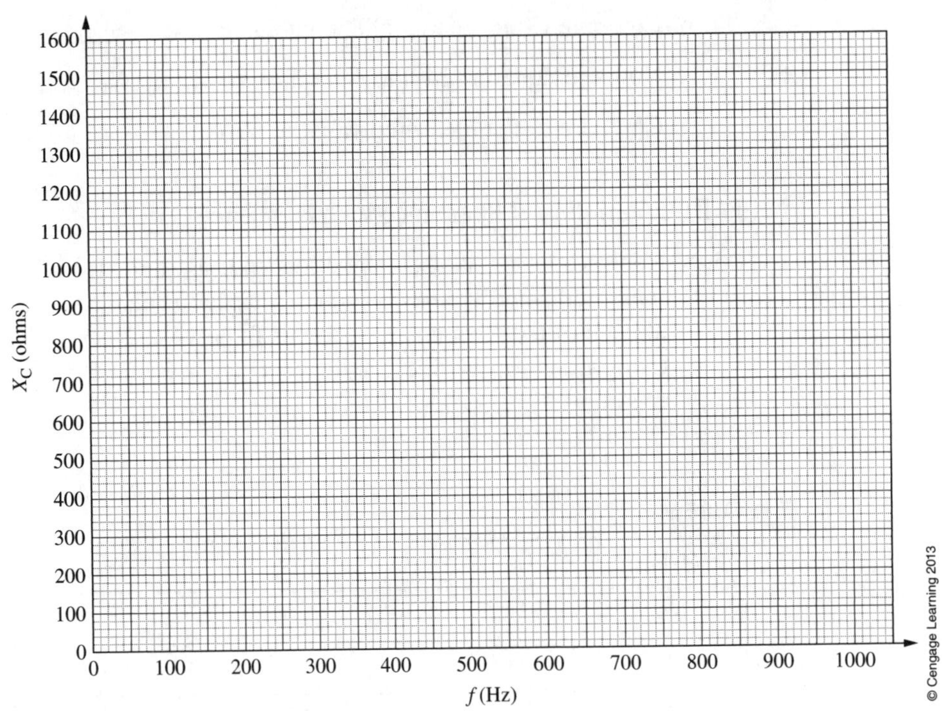

FIGURE 14-3 Plot of capacitive reactance versus frequency.

COMPUTER ANALYSIS

6. Using either Multisim or PSpice, build a circuit on your screen consisting of an ac source and a 1-μF capacitor, then compute and plot its reactance from $f = 100$ Hz to 1 kHz. (The source voltage can be anything, for example, 10 V.)
   a. For PSpice, see Example 16-19 of the textbook as reference.
   b. For Multisim 11, click *Simulate, Analyses,* then *AC Analysis*. When the dialog box opens, choose *Linear* for both *Sweep type* and *Vertical scale*. Click the *Output* tab and create the expression *V(1)/I(C1),* then click *Simulate*. Compare the magnitude plot against Figure 16-36 of the textbook.
   c. For Multisim 9, click *Simulate, Analyses,* then *AC Analysis*. When the dialog box opens, choose *Linear* for both *Sweep type* and *Vertical scale*. Click the *Output* tab and create the expression *$1*/vv1#branch (you will see these variables in your dialog box), then click *Simulate*. Compare the magnitude plot against Figure 16-36 of the textbook. (Ignore the phase plot—it should be a straight line at −90°, but Multisim 9 fouls it up. However, Multisim 11 does it correctly.)

PROBLEM

7. Consider an ideal capacitor. If you double the capacitance and triple the frequency, what happens to the current if the applied sinusoidal voltage remains the same?

Name ______________________

Date ______________________

Class ______________________

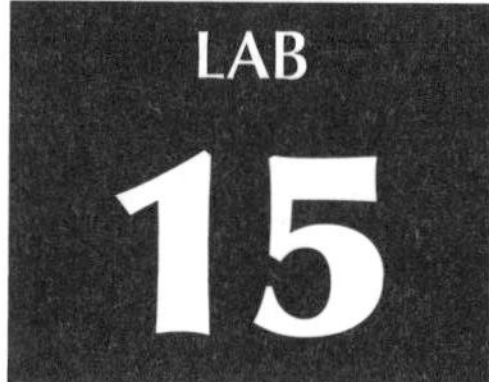

Inductive Reactance

OBJECTIVES

After completing this lab, you will be able to
- measure phase difference between voltage and current in an inductance,
- measure inductive reactance and verify theoretically,
- determine the effect of frequency on inductive reactance.

EQUIPMENT REQUIRED

☐ Dual channel oscilloscope
☐ Signal or function generator
☐ Digital multimeter (DMM) (two)

COMPONENTS

☐ Resistors: 10-Ω, 100-Ω, 1/4-W
☐ Inductors: 2.4-mH, two required. (Hammond Part #1534 or equal. This inductor has a very small resistance which is necessary to approximate an ideal inductor.)

PRE-STUDY (OPTIONAL)

Interactive simulation *CircuitSim 16-3* may be used as pre-study for this lab. From the Student Premium Website, select *Learning Circuit Theory Interactively on Your Computer,* download and unzip *Ch 16* (if you have not already done so), and run the corresponding *Sim* file. (Note that its circuits are similar to but not identical to those of this lab.)

TABLE 15-1

Instrument	Manufacturer/Model No.	Serial No.
DMM #1		
DMM #2		
Dual channel oscilloscope		
Signal or function generator		

TEXT REFERENCE

Section 16.5 INDUCTANCE AND SINUSOIDAL ac

DISCUSSION

Inductance and Sinusoidal ac. When an ideal inductance is connected to a sinusoidal voltage source, current lags voltage by 90° as illustrated in Figure 15-1. Thus, if

$$v_L = V_m \sin \omega t \tag{15-1}$$

then

$$i_L = I_m \sin (\omega t - 90°) \tag{15-2}$$

where

$$I_m = V_m/X_L \tag{15-3}$$

The quantity X_L is termed *inductive reactance* and is given by the formula

$$X_L = \omega L\ \Omega \tag{15-4}$$

where $\omega = 2\pi f$ rad/s. Inductive reactance represents the opposition that the inductance presents to current and is directly proportional to the product of inductance and frequency. Thus, the higher the frequency, the greater the opposition.

Effective (rms) Voltage and Current Relationships. In practice, we usually use effective values rather than peak values. Their ratios are the same, however. Thus, reactance can also be expressed as

$$X_L = V_L/I_L\ \Omega \tag{15-5}$$

where V_L and I_L are the rms values of v_L and i_L, respectively. (Note that V_L and I_L are the values that you read on meters, while V_m and I_m are the values that you read on the oscilloscope.)

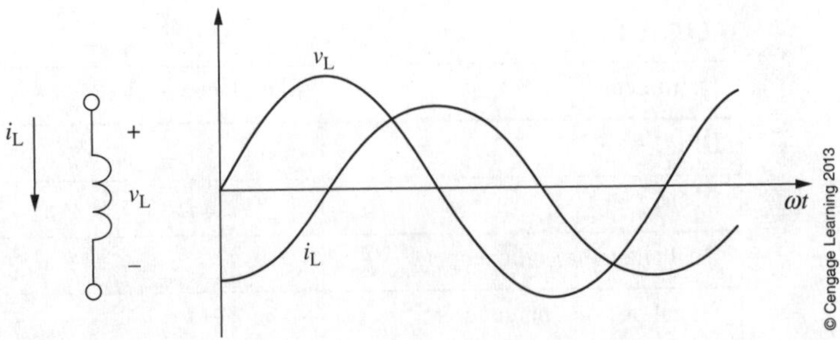

FIGURE 15-1 Voltage and current for an inductance.

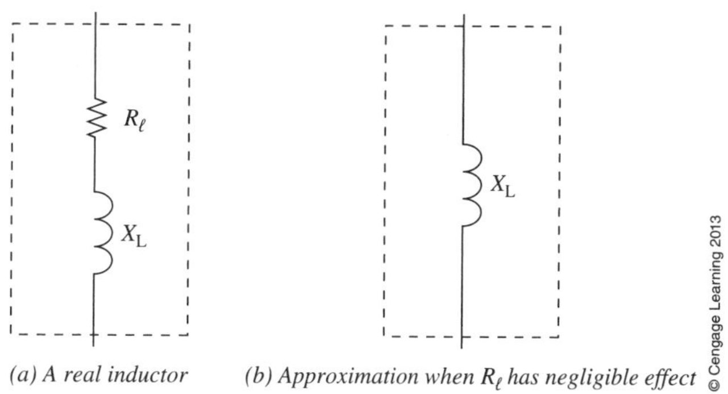

FIGURE 15-2 The ideal inductor approximation.

Practical Inductors. In reality, inductors have resistance as well as inductance—see Figure 15-2(a). However, if X_L is large compared to R_ℓ, we can neglect resistance and use the ideal model of (b). This means the simple formulas above apply. (In this lab, we will use an inductance where this approximation is very good at higher frequencies. However, at lower frequencies, the error in the approximation starts to show up. You will have a chance to see how good the approximation is at these lower frequencies.)

MEASUREMENTS

PART A: Phase Relationships for Inductance

Consider Figure 15-3. Voltage v_S across R_S is in phase with inductor current i_L. Provided circuit resistance is small compared to X_L, source voltage is approximately equal to the v_L. Thus, the phase difference between Ch1 and Ch2 is approximately equal to the angle between v_L and i_L. It should be close to 90°.

1. a. Measure coil resistance and inductance and the resistance of the sensing resistor, then assemble the circuit of Figure 15-3.

   $L =$ ___________ $R_\ell =$ ___________ $R_S =$ ___________

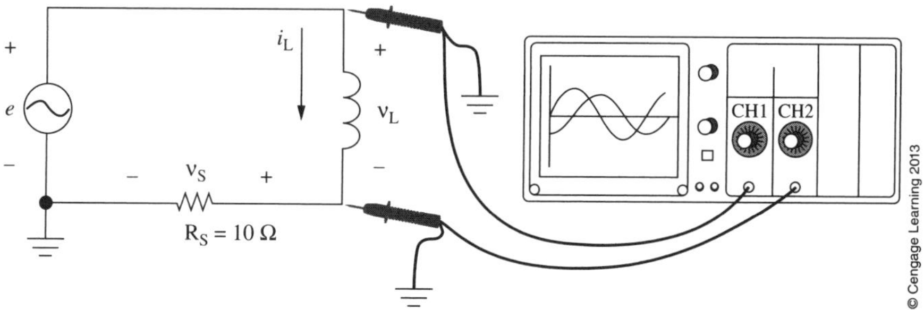

FIGURE 15-3 Circuit resistance is very small. Therefore, $v_L \approx$ e.

b. Use a 40-kHz sine wave. Sketch the waveforms below. Label Ch1 as v_L and Ch2 as i_L. Using the procedure of Lab 13, determine the phase shift and indicate it on the diagram. How close is it to the theoretical value of 90°? Vary the frequency a few kHz and observe the shift. Is there any change? Based on this observation, state in your own words the phase relationship between voltage and current for an inductance.

c. The inductor will behave approximately as an ideal inductor provided $X_L >> R$ where $R = R_\ell + R_s$. Check the approximation at 40 kHz. Use the measured R_ℓ and L determined in Test 1(a).

Practical Notes

1. For this lab, you need to measure voltages with frequencies up to 10 kHz, so ensure that your DMMs can handle this range. If they cannot, use VOMs (most VOMs can), or if you do not have suitable meters, make measurements with the oscilloscope.
2. For this lab, you need to be able to set frequency quite accurately. However, the frequency dial on some signal generators is not accurately calibrated. If you have such a generator, use your oscilloscope to set frequency.

PART B: Magnitude Relationships for Inductance

2. a. Replace resistor R_S of Figure 15-3 with an accurately measured 100-Ω resistor.

R_S = ___________

b. Set the signal generator to a 10-kHz sine wave, 6 V peak to peak. Measure voltage V_L across the inductor and voltage V_S across R_S. Calculate current I_L using Ohm's law and the measured value of V_S.

V_L = ___________ V_S = ___________ I_L = ___________

c. Calculate the reactance of the inductor using the measured V_L and I_L.

$X_{L(measured)} = V_L/I_L =$ ____________

d. If possible, measure L using a bridge or *RLC* meter. (Otherwise, use the nominal value.) Compute X_L using Equation 15-4. Compare to the measured value of Test 2(c).

$L =$ ____________ $X_{L(computed)} =$ ____________

3. a. Add the second inductor in series and determine total reactance using the same procedure that you used in Test 2(b) and (c).

$V_L =$ ____________ $V_S =$ ____________

$I_L =$ ____________ $X_{T(measured)} =$ ____________

b. If possible, measure L for the second inductor and compute L_T for the series combination. $L_T =$ ____________. Now use Equation 15-4 to determine $X_{T(computed)}$. $X_{T(computed)} =$ ____________. Compare to the measured value of Test 3(a).

4. a. Connect both inductors in parallel and determine their equivalent reactance using the procedure that you used in Test 2(b) and (c).

V_L = ____________ V_S = ____________

I_L = ____________ $X_{eq(measured)}$ = ____________

b. Compute L_{eq} for the parallel combination. L_{eq} = ____________. Now use Equation 15-4 to determine $X_{eq(computed)}$. $X_{eq(computed)}$ = ____________. Compare to the measured value of Test 4(a).

PART C: Variation of Reactance with Frequency

5. a. Using one of the 2.4-mH inductors, measure V_L and V_S and compute reactance (using the procedure of Test 2) at each of the frequencies listed in Table 15-2. Record as $X_{L(measured)}$.
   b. Using Equation 15-4, compute reactance at each frequency and record as $X_{L(computed)}$.
   c. Using the graph sheet of Figure 15-4, plot measured and computed reactances versus frequency. Label each plot.
   d. The plots should pass through $X_L = 0\ \Omega$ at $f = 0$ Hz. Do they? (Extrapolate both plots to find out.)
   e. Analyze the results. That is, comment on how well the measured values agree with the computed values. Is the agreement poorer as the frequency gets lower? If so, why?

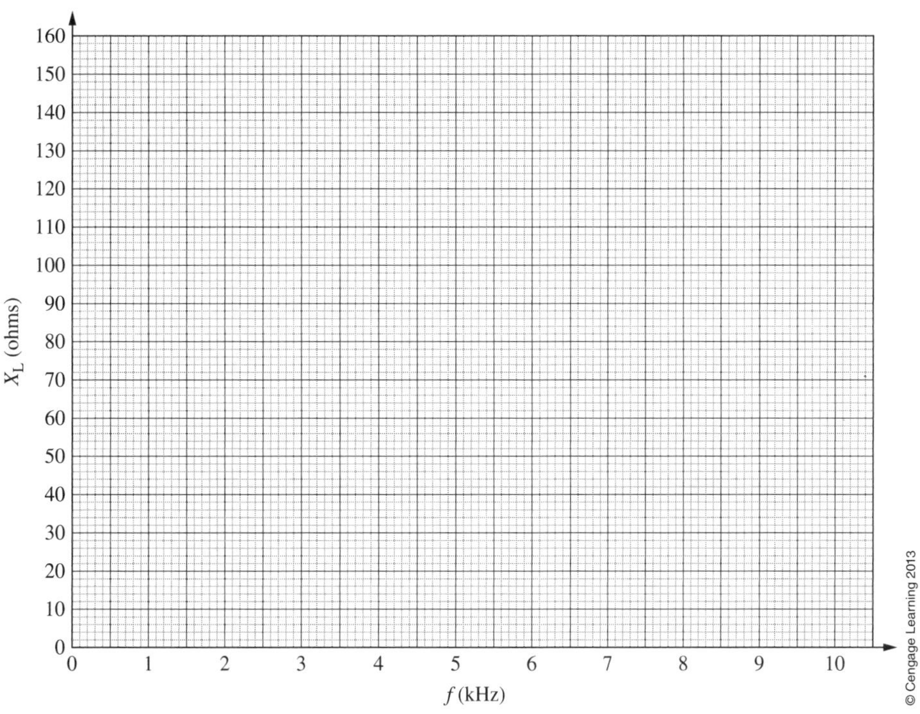

FIGURE 15-4 Plot of inductive reactance versus frequency.

TABLE 15-2

f(kHz)	V_L	V_S	I_L	$X_{L(measured)}$	$X_{L(computed)}$
1					
2					
3					
4					
5					
6					
7					
8					
9					
10					

PROBLEM

6. Considering ideal inductors, if you double inductance and triple frequency, what happens to current if the applied sinusoidal voltage remains the same?

FOR FURTHER INVESTIGATION AND DISCUSSION

Analysis Notes

1. Normally frequency plots are made with a log scale for frequency. However, we want to investigate the linearity of impedance versus frequency—thus, use a linear scale for both the horizontal and vertical axes of your plot.
2. See Lab 14, Part 6 for hints on using Multisim and PSpice for this kind of analysis.

How well do ideal inductors model real inductors? Investigate this question by comparing the impedance characteristic of an ideal inductor to the more realistic model depicted in Section 13.6, Figure 13-21(a) of the text. Assume a 2.4-mH inductor with resistance of 25 Ω and stray capacitance of 0.5 nF. To aid your discussion, do the following.

a. Using Multisim or PSpice, compute and plot the magnitude of the impedance (i.e., the reactance) of the ideal inductor from $f = 1$ Hz to 10 kHz.
b. Repeat using the model of Figure 13-21(a). Run two sets of plots, one from $f = 1$ Hz to $f = 10$ kHz, and the other from $f = 1$ Hz to $f = 4$ kHz (to better see the low-frequency effects).

Write a short report discussing your findings.

Name ______________________

Date ______________________

Class ______________________

Power in ac Circuits

OBJECTIVES

After completing this lab, you will be able to
- measure power in a single-phase circuit,
- verify power relationships,
- verify power factor relationships,
- determine the effect of adding power factor correction.

EQUIPMENT REQUIRED

- ☐ Single-phase wattmeter
- ☐ ac ammeter
- ☐ Digital multimeter (DMM)

COMPONENTS

☐ Resistor:	100-Ω, rated 200 W
☐ Capacitor:	30-μF, nonelectrolytic, rated for operation at 120 Vac (1 required)
☐ Inductor:	Approximately 0.2 H, rated to handle 2 amps

Safety Note

In this lab, you will be working with 120 Vac. This voltage is dangerous, and you must be aware of and observe safety precautions. Familiarize yourself with your laboratory's safety features. Observe all safety instructions posted and instructions from your lab supervisor. **Ensure that power is off when you are assembling, changing, or otherwise working on your circuit.**

PRE-STUDY (OPTIONAL)

Interactive simulations *CircuitSim 17-1 and 17-9* may be used as pre-study. From the Student Premium Website, select *Learning Circuit Theory Interactively on Your Computer,* download and unzip *Ch 17* (if you have not already done so), and run the corresponding *Sim* files. (Note that their circuits are similar to but not identical to those of this lab.)

EQUIPMENT USED

TABLE 16-1

Instrument	Manufacturer/Model No.	Serial No.
Single-phase wattmeter		
ac Ammeter		
DMM		

TEXT REFERENCE

Section 17.7 THE RELATIONSHIP BETWEEN *P*, *Q*, AND *S*
Section 17.8 POWER FACTOR
Section 17.9 ac POWER MEASUREMENT

DISCUSSION

Power in ac Systems. Power to an ac load is given by

$$P = V I \cos \theta \quad (16\text{-}1)$$

where *V* and *I* are the magnitudes of the rms load voltage and current, respectively, and θ is the angle between them. (θ is the angle of the load impedance.) The power factor of the load is

$$F_p = \cos \theta \quad (16\text{-}2)$$

> **Note**
>
> Some wattmeters have the voltage sensing circuit connected internally, leaving only three external connection terminals—see Figure 17-26 of the textbook.

Measuring Power in ac Circuits. Power is measured as in Figure 16-1(a) or (b) with the wattmeter connected so that current passes through its current coil CC, and load voltage is applied to its voltage-sensing circuit. For this lab, either connection will work—see Note.

Power Factor Correction. If a load such as that shown in Figure 16-1 has poor power factor (i.e., θ is large), it will draw excessive current relative to the power transferred to it from the source. One way to improve the power flow in the system is to add power factor correction at the load. Since most power system loads

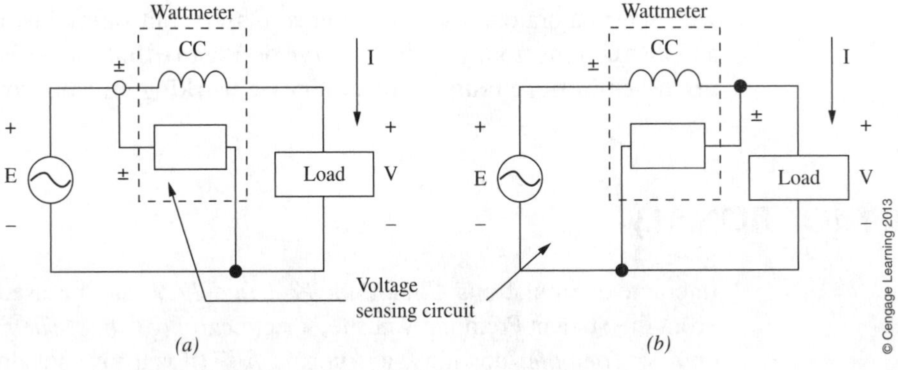

FIGURE 16-1 Measuring power in an ac circuit. We usually simplify the wattmeter representation as in Figure 16-2.

are inductive (because they contain inductive elements such as electric motors), this may be done by adding capacitors in parallel across the load. For loads with a poor power factor, this can dramatically reduce source current. It is important to note however that power factor correction does not change the power requirements of the load—it simply supplies the needed reactive power locally, rather than from the source. However, it greatly reduces source current.

MEASUREMENTS

If possible, use a variable ac power supply (such as a variable autotransformer) as the source—see General Safety Notes.

General Safety Notes

1. With power off, assemble your circuit and double-check it.
2. Have your instructor check the circuit before you energize it. If you are using a Powerstat or other variable ac source, gradually increase voltage from zero, watching the meters for signs of trouble.
3. Turn power off before changing your circuit for the next test.

Check with your instructor for specific safety instructions.

PART A: Power in a Purely Resistive Circuit

1. Measure the 100-Ω resistor. R = __________. Connect the circuit as in Figure 16-2. (The ammeter may be a standard ac ammeter or the ac current range of a DMM. For the load resistance used here, a 2-A range is adequate. Select a wattmeter to match—for example, a wattmeter with approximately a 300-W scale.) Carefully measure load voltage, current, and power and record in Table 16-2.
2. Using the values of V and R measured in Test 1, compute current, then determine power to the load using each of the formulas $P = VI$, $P = I^2R$, and $P = V^2/R$. Compare to the value measured with the wattmeter.

PART B: Power to a Reactive Load

3. a. De-energize the circuit and add 30 μF of capacitance in parallel with the load as in Figure 16-3. Measure load voltage, current, and power and record in Table 16-3.

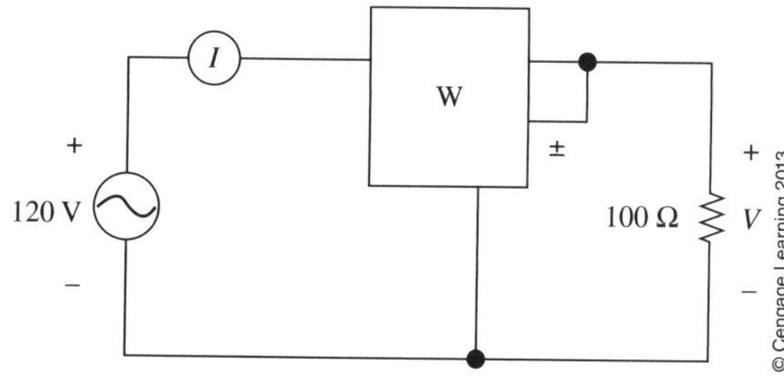

FIGURE 16-2 Purely resistive load. (The ± voltage connection may be internal to the wattmeter.)

TABLE 16-2

	Measured Values
V	
I	
P	

© Cengage Learning 2013

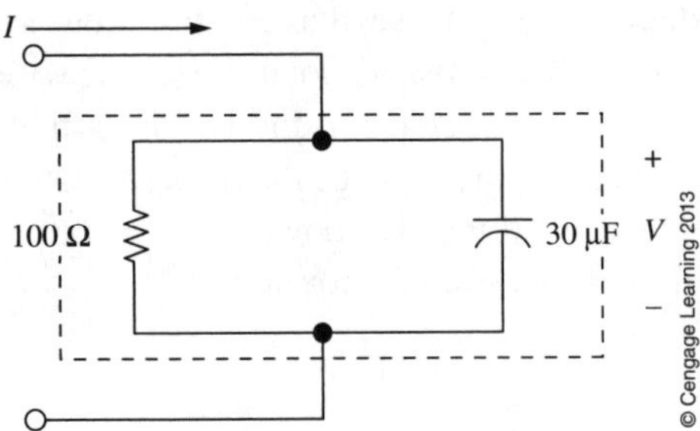

FIGURE 16-3 Leading power factor load.

TABLE 16-3

	Measured Values
V	
I	
P	

b. Calculate real power using the measured voltage and resistance and compare to the measured value.

c. Calculate reactive power Q.

d. Draw the power triangle using the values computed in Tests 3(b) and 3(c). Using the apparent power S from this triangle, compute current. Calculate the percent difference between this and the value measured in Test 3(a). What is the likely source of difference?

e. Determine the power factor from the data of Table 16-3 and Equation 16-1.

f. Determine the power factor from the power triangle of Test 3(d).

g. Determine the load impedance **Z** from Figure 16-3, then calculate power factor using Equation 16-2.

h. Compare the values of power factor determined in parts (e), (f), and (g)—that is, comment on any differences and why they might have occurred.

TABLE 16-4

Measured Values	
V	
I	
P	

© Cengage Learning 2013

PART C: Power Factor Correction

4. a. Replace the load with the 0.2-H coil, Figure 16-4(a). (Measure its resistance and inductance first.) Measure load voltage, current, and power, and record in Table 16-4.
   b. Add the 30 μF as in Figure 16-4(b) and observe the decrease in current. Using the techniques of Chapter 17, solve for source current and compare.

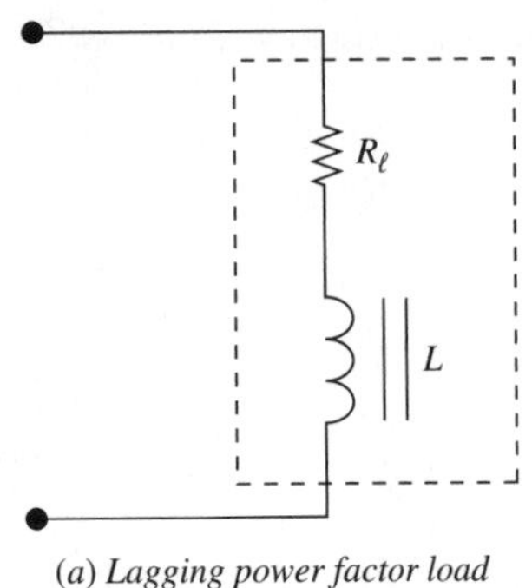

(a) Lagging power factor load

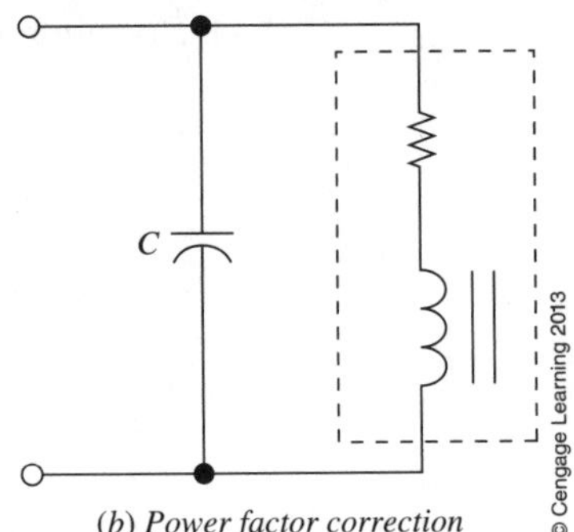

(b) Power factor correction

FIGURE 16-4

COMPUTER ANALYSIS

5. Consider Figure 17-22 of the text. Using your calculator, convert the plant load to an equivalent resistance and inductance in series. Omitting the capacitor, set up the resulting circuit using either Multisim or PSpice.
   a. Insert an ammeter (IPRINT if you are using PSpice) and configure for ac operation. Set the source frequency to 60Hz and determine the magnitude of the line current.
   b. Add the capacitor across the load and again determine the magnitude of the line current.
   c. Change the capacitor to the value that yields unity power factor (*see Practice Problem 4 of the textbook*) and again measure current.

   How do the results compare to those in the text?

PROBLEMS

6. A motor delivers 50 hp to a load. The efficiency of the motor is 87% and its power factor is 69%. The source voltage is 277 V, 60 Hz.
   a. Determine the power input to the motor.
   b. Draw the power triangle for the motor.
   c. Determine the source current.
   d. Determine how much capacitance is required to correct the power factor to unity.
   e. What is the source current when the capacitance of (c) is added?
   f. If 50% more capacitance than computed in (c) is added, determine the source current.

Name ______________________

Date ______________________

Class ______________________

Series ac Circuits

OBJECTIVES

After completing this lab, you will be able to

- calculate current and voltages for a simple series ac circuit,
- measure voltage magnitude and phase angle in a simple series ac circuit,
- verify Kirchhoff's voltage law using measured results,
- measure the internal impedance of a sinusoidal voltage source.

EQUIPMENT REQUIRED

☐ Dual-trace oscilloscope
☐ Signal generator (sinusoidal function generator)
Note: Record this equipment in Table 17-1.

COMPONENTS

☐ Resistors: 680-Ω (1/4-W carbon, 5% tolerance)
☐ Capacitors: 0.22-μF (10% tolerance)

EQUIPMENT USED

TABLE 17-1

Instrument	Manufacturer/Model No.	Serial No.
Oscilloscope		
Signal generator		

© Cengage Learning 2013

TEXT REFERENCE

Section 18.1 OHM'S LAW FOR ac CIRCUITS
Section 18.2 ac SERIES CIRCUITS
Section 18.3 KIRCHHOFF'S VOLTAGE LAW AND THE VOLTAGE DIVIDER RULE

DISCUSSION

Series ac circuits behave in a manner similar to the operation of dc circuits. Ohm's law, Kirchhoff's current and voltage laws, and the various circuit analysis rules apply for ac circuits as well as for dc circuits. The principal differences in how these rules and laws apply to ac circuits are outlined as follows:

- All impedances are complex numbers and may be expressed either in rectangular form or polar form (e.g., $\mathbf{Z} = 10\ \Omega + j20\ \Omega = 22.36\ \Omega\angle 63.43°$). The real component of the impedance vector represents the resistance while, the imaginary component corresponds to the reactance. The imaginary component of the impedance will be positive if the reactance is inductive and negative if the reactance is capacitive.
- *Time-domain values* such as $v_C = 2\sin(\omega t + 30°)$ are converted into *phasor domain* ($\mathbf{V}_C = 1.414\ \text{V}\angle 30°$) to permit arithmetic operations using complex numbers.
- Although time domain values are always expressed using peak values, phasor voltages and currents are always expressed with magnitudes in rms (root-mean-square).
- Power calculations are performed using rms values and must consider the *power factor* of the circuit or component.

CALCULATIONS

1. Refer to the circuit of Figure 17-1. Determine the reactance of the capacitor at a frequency of $f = 2$ kHz. Express the circuit impedance $\mathbf{Z}_T$ in both the rectangular form and the polar form.

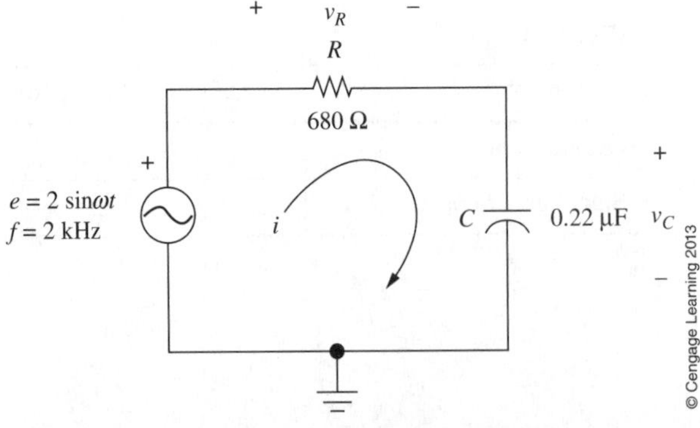

FIGURE 17-1 Series ac circuit.

X_C	
Z_T	

2. Convert the time domain form of the voltage source of the circuit of Figure 17-1 into its equivalent phasor domain form. Calculate the phasor current **I** and solve for the phasor voltages $\mathbf{V}_R$ and $\mathbf{V}_C$. Enter your results in Table 17-2.

TABLE 17-2

E	
I	
$\mathbf{V}_R$	
$\mathbf{V}_C$	

3. Use complex algebra together with the phasor forms of the voltages (**E**, $\mathbf{V}_R$, and $\mathbf{V}_C$) to verify that Kirchhoff's voltage law applies.

$$\sum \mathbf{V} = 0 = \mathbf{E} - \mathbf{V}_R - \mathbf{V}_C \qquad (17\text{-}1)$$

4. Convert the phasor forms of **I,** $\mathbf{V}_R$, and $\mathbf{V}_C$ into their equivalent time domain values. Enter the results in Table 17-3.

TABLE 17-3

i	
v_R	
v_C	

MEASUREMENTS

5. Assemble the circuit of Figure 17-1.
6. Refer to the pictorial diagram of Figure 17-2. Connect Ch1 of the oscilloscope to the output of the signal generator. Set the oscilloscope to have an automatic sweep and use Ch1 as the trigger source. Adjust the output of the generator to provide a sinusoidal voltage with an amplitude of 2.0 V (4.0 $V_{p\text{-}p}$) and a frequency of $f = 2$ kHz ($T = 500$ μs).
7. Use Ch2 of the oscilloscope to observe the signal across the capacitor $\mathbf{V}_C$. (You should still have the output of the signal generator displayed on Ch1.) Notice that the voltage across the capacitor is lagging the generator voltage. Adjust the time/division to provide at least one full cycle on the oscilloscope. Accurately sketch and label both the generator voltage $\boldsymbol{e}$ and the capacitor voltage v_C in the space provided on Graph 17-1. In the space below, record the volts/division setting for each channel and the time/division of the oscilloscope.

Ch1: __________ V/Div

Ch2: __________ V/Div

Time: __________ μs/Div

8. Use the following expression to calculate the phase angle between capacitor voltage and the generator voltage. (The phase angle is negative since v_C lags $\boldsymbol{e}$.)

$$\theta = \frac{\Delta t}{T} \times 360° \qquad (17\text{-}2)$$

θ_1	

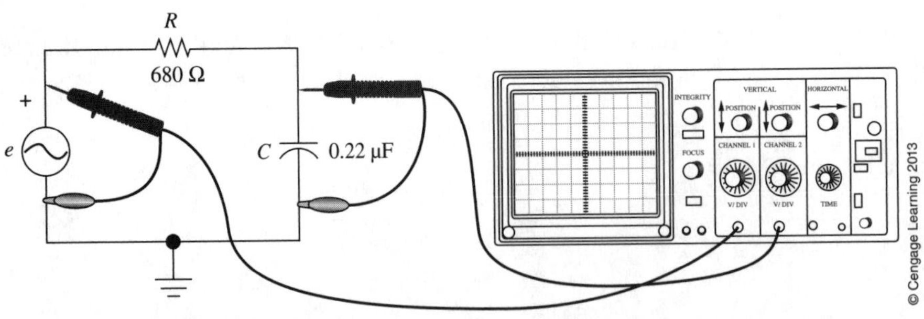

FIGURE 17-2 Equipment connection for voltage measurement.

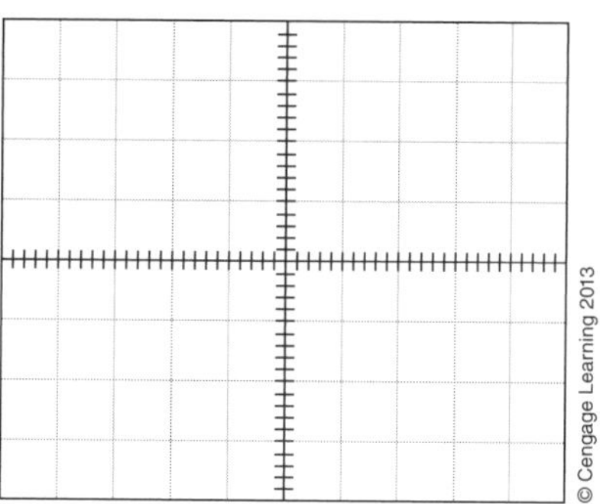

GRAPH 17-1

9. It is not possible to measure the voltage across the resistor directly since the internal ground connection in the probe of the oscilloscope will result in a short circuit across the capacitor. To overcome this problem, switch the oscilloscope to show the *difference mode* (Ch1–Ch2), making certain that both Ch1 and Ch2 of the oscilloscope are on the same VOLTS/DIV setting. You should observe a single display, corresponding to the resistor voltage v_R. Since you are using the Ch1 signal as the trigger source, the observed waveform provides the phase shift between the signal generator and the resistor voltage. **Be careful not to adjust the horizontal position. If you accidently move the display, you will need to readjust the display to obtain the waveforms of Step 7.** You should observe that the resistor voltage is leading the generator voltage. Sketch and label the observed resistor voltage v_R as part of Graph 17-1.
10. Calculate the phase shift between the resistor voltage and the generator voltage. Since the resistor voltage is leading the generator voltage, the phase angle is positive.

θ_2	

Internal Impedance of Voltage Sources

All signal generators have some internal impedance, which tends to reduce the voltage between the terminals of the signal generator when the generator is under load. Most low-frequency signal generators have a nominal internal impedance of 600 Ω. Other signal generators may have impedance values of 50 Ω, 75 Ω, or 300 Ω, depending on the application. Figure 17-3 shows a simple representation of the output of a signal generator.

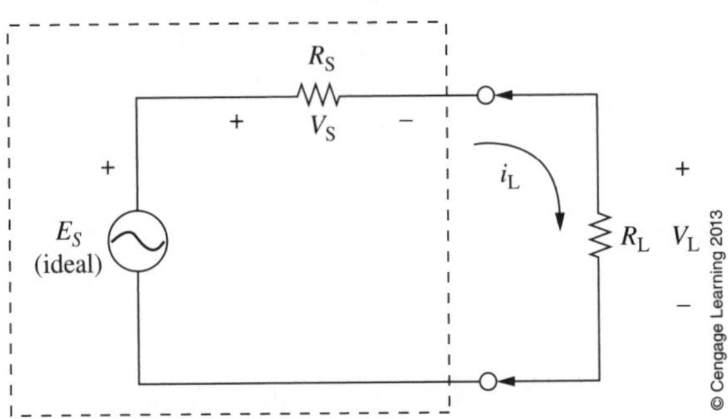

FIGURE 17-3 Internal impedance of a signal generator.

11. Connect the oscilloscope between the output terminals of the signal generator. Adjust the generator to provide a sinusoidal output having an amplitude of 1 V (appearing as 2 $V_{p\text{-}p}$) and a frequency of 1 kHz. Since the signal generator is not loaded, this voltage represents the signal of the ideal voltage source $E_S = 2\ V_{p\text{-}p}$.
12. Connect a 680-Ω load resistor between the output terminals of the signal generator. Reconnect the oscilloscope between the output terminals of the signal generator. You should see that the voltage at the output has now decreased significantly. This represents the loaded output voltage. Measure and record the loaded peak-to-peak output voltage V_L.

V_L	$V_{p\text{-}p}$

CONCLUSIONS

13. Refer to Graph 17-1. Calculate the amplitude V_R of the resistor voltage and the amplitude V_C of the capacitor voltage. From Steps 8 and 10, evaluate the phase angle (with respect to the generator voltage ***e***) for each of these voltages. Express each voltage in its time domain form [e.g., $v_C = V_C \sin(\omega t + \theta_1)$]. Convert the amplitudes into rms quantities and express each voltage in its phasor form. Enter all results in Table 17-4.

TABLE 17-4

Resistor		Capacitor	
V_R	θ	V_C	ρ
$v_R =$		$v_C =$	
$\mathbf{V}_R =$		$\mathbf{V}_C =$	

14. Compare the measured sinusoidal capacitor voltage v_C of Table 17-4 to the theoretical value of Table 17-2.

____

____

15. Compare the measured sinusoidal resistor voltage v_R of Table 17-4 to the theoretical value of Table 17-2.

____

____

16. Calculate the actual signal generator resistance using Ohm's law and the measurement of Step 12.

R_S	

FOR FURTHER INVESTIGATION AND DISCUSSION

Use Multisim or PSpice to simulate the circuit of Figure 17-1. Measure the amplitude and phase angles of $\boldsymbol{V}_R$ and $\boldsymbol{V}_C$. Compare these values to the actual observations. Explain any discrepancies.

Name ______________________

Date ______________________

Class ______________________

LAB 18

Parallel ac Circuits

OBJECTIVES

After completing this lab, you will be able to

- measure voltages in a parallel circuit using an oscilloscope,
- use an oscilloscope to indirectly measure current magnitude and phase angles in a simple parallel ac circuit,
- compare measured values to theoretical calculations and verify Kirchhoff's current law,
- determine the power dissipated by a parallel ac circuit.

EQUIPMENT REQUIRED

☐ Dual-trace oscilloscope
☐ Signal generator (sinusoidal function generator)
Note: Record this equipment in Table 18-1.

COMPONENTS

☐ Resistors: 10-Ω (3), 470-Ω (1/4-W carbon, 5% tolerance)
☐ Capacitors: 3300-pF (10% tolerance)
☐ Inductors: 1-mH (iron core, 5% tolerance—Hammond 1534A or equivalent)

EQUIPMENT USED

TABLE 18-1

Instrument	Manufacturer/Model No.	Serial No.
Oscilloscope		
Signal generator		

TEXT REFERENCE

Section 18.4 ac PARALLEL CIRCUITS
Section 18.5 KIRCHHOFF'S CURRENT LAW AND THE CURRENT DIVIDER RULE

DISCUSSION

The equivalent impedance of a parallel circuit is determined by finding the sum of the admittances of all branches. The important point to remember when calculating total admittance is that all admittances are expressed as complex values. This means that we must use complex algebra to find the solution, which is also a complex number.

Refer to the circuit of Figure 18-1. The equivalent admittance of the circuit is determined as

$$\mathbf{Y}_T = \frac{1}{R} + j\frac{1}{X_C} - j\frac{1}{X_L} \tag{18-1}$$

This gives an equivalent circuit impedance of

$$\mathbf{Z}_T = \frac{1}{\mathbf{Y}_T} = \frac{1}{\frac{1}{R} + j\frac{1}{X_C} - j\frac{1}{X_L}} \tag{18-2}$$

In order to further analyze the circuit, it is necessary to convert the ac voltage source into its equivalent phasor form. The circuit current **I** is then easily determined by applying Ohm's law. The current through each component in the circuit is similarly found by applying Ohm's law to each branch or by applying the current divider rule as follows:

$$\mathbf{I}_x = \frac{\mathbf{E}}{\mathbf{Z}_x} = \frac{\mathbf{Z}_T}{\mathbf{Z}_x}\mathbf{I} \tag{18-3}$$

Regardless of the method used to determine currents, Kirchhoff's current law must apply at any node in the circuit. Therefore,

$$\sum \mathbf{I} = 0 \tag{18-4}$$

where each current is in its phasor form.

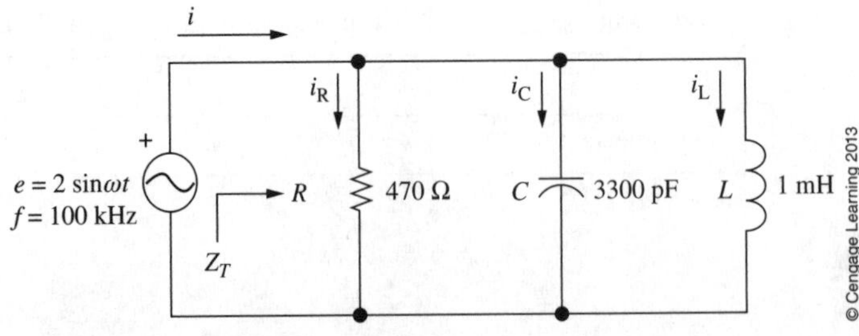

FIGURE 18-1 Parallel ac circuit.

CALCULATIONS

1. Refer to the circuit of Figure 18-1. Determine the reactances of the capacitor and the inductor at a frequency of $f = 100$ kHz. Calculate the circuit impedance $\mathbf{Z}_T$ and express the result in both rectangular and polar form. Enter the data in Table 18-2.

TABLE 18-2

X_C	
X_L	
$\mathbf{Z}_T$	

2. Convert the time domain form of the voltage source of the circuit of Figure 18-1 into its equivalent phasor domain form. Calculate the phasor current **I** and solve for the phasor currents $\mathbf{I}_R$, $\mathbf{I}_C$, and $\mathbf{I}_L$. Enter your results in Table 18-3.

TABLE 18-3

$\mathbf{E}$	
$\mathbf{I}$	
$\mathbf{I}_R$	
$\mathbf{I}_C$	
$\mathbf{I}_L$	

3. Use complex algebra together with the phasor currents $\mathbf{I}_R$, $\mathbf{I}_C$, and $\mathbf{I}_L$ to verify that Kirchhoff's current law applies at node *a*.

$$\mathbf{I} = \mathbf{I}_R + \mathbf{I}_C + \mathbf{I}_L \tag{18-5}$$

4. Convert the phasor currents **I**, $\mathbf{I}_R$, $\mathbf{I}_C$, and $\mathbf{I}_L$ into their equivalent time domain forms. Enter the results in Table 18-4.

TABLE 18-4

i	
i_R	
i_C	
i_L	

Current cannot be measured directly with an oscilloscope. However, by strategically placing small series *sensing resistors* into a circuit and then measuring voltage across these resistors, current through each branch is found by applying Ohm's law.

MEASUREMENTS

5. Assemble the circuit of Figure 18-2. Notice that three 10-Ω *sensing resistors* have been added into the circuit to help in determining branch currents. Since these resistors are small in comparison to the impedance in the branch, they will not significantly load the circuit, and their effects may be ignored.
6. Connect Ch1 of the oscilloscope to the output of the signal generator at point *a*. Set the oscilloscope to have an automatic sweep and use Ch1 as the *trigger source*. Adjust the output of the generator to provide a sinusoidal voltage with an amplitude of 2.0 V (4.0 $V_{p\text{-}p}$) and a frequency of $f = 100$ kHz ($T = 10$ μs).
7. Use Ch2 of the oscilloscope to observe the voltage at point *b*. This is the voltage across sensing resistor R_1. Measure the observed peak-to-peak voltage V_1. Measure the phase angle θ_1 of the voltage V_1 with respect to the generator voltage. Record your results in Table 18-5.

   Since the observed waveform is across the 10-Ω resistor, the amplitude of the current *i* is now easily calculated from the peak-to-peak voltage as

$$I = \frac{V_1/2}{10\ \Omega} \qquad (18\text{-}6)$$

Write the time-domain expression for *i* using your measurements and calculations. Enter your results in Table 18-5.

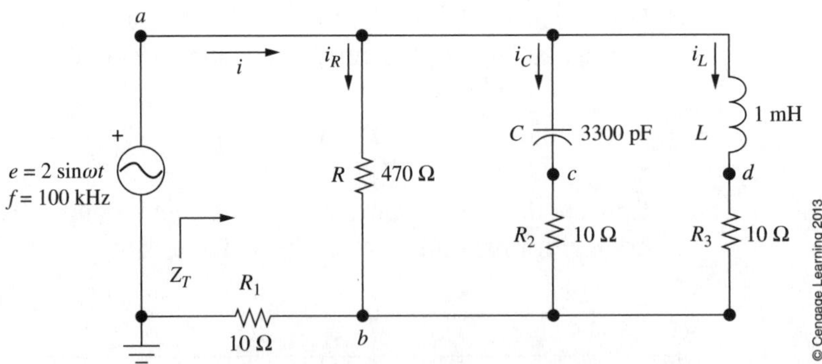

FIGURE 18-2 Using sensing resistors for current measurement.

TABLE 18-5

V_1	$V_{p\text{-}p}$
θ_1	
i	

8. Since we no longer need R_1 in the circuit, it may be removed and replaced with a short circuit. Measure the peak-to-peak voltage V_R and calculate current i_R using Ohm's law. Use your measurements and calculations to record the time-domain expression for i_R in Table 18-6.

TABLE 18-6

V_R	$V_{p\text{-}p}$
θ	
i_R	

9. Ensure that Ch1 of the oscilloscope is connected to point a (generator voltage) and that the sensing resistor R_1 is removed from the circuit.
   Connect Ch2 of the oscilloscope to point c to measure the voltage across the sensing resistor R_2. Record the peak-to-peak voltage V_2. Measure and record the phase angle between the generator voltage and the sensing voltage V_2. You should observe that the voltage V_2 (and hence, the current) is leading the generator voltage. Use your measurements and Ohm's law to determine the sinusoidal expression for i_C. Record the result in Table 18-7.

TABLE 18-7

V_1	$V_{p\text{-}p}$
θ_2	
i_C	

10. With Ch1 of the oscilloscope connected to point a, connect Ch2 of the oscilloscope to point c. Measure and record the peak-to-peak voltage across the sensing resistor R_3. Measure and record the phase angle between the generator voltage and the sensing voltage V_2. You should observe that the voltage V_3 (and hence, the current) is lagging the generator voltage. Use your measurements and Ohm's law to determine the sinusoidal expression for i_L. Record the result in Table 18-8.

TABLE 18-8

V_3	$V_{p\text{-}p}$
θ_3	
i_L	

CONCLUSIONS

11. Compare the measured currents i, i_R, i_C, and i_L of Table 18-5 through Table 18-8 to the theoretical values of Table 18-4.

____

____

____

12. Convert each of the measured currents i, i_R, i_C, and i_L into its phasor form. Enter the results in Table 18-9.

TABLE 18-9

$\mathbf{I}$	
$\mathbf{I}_R$	
$\mathbf{I}_C$	
$\mathbf{I}_L$	

13. Use complex algebra to show that the data of Table 18-9 verifies Kirchhoff's current law.

14. Use the measured currents $\mathbf{I}$, $\mathbf{I}_R$, $\mathbf{I}_C$, and $\mathbf{I}_L$ to calculate the total power dissipated by the circuit in Figure 18-2. (You should observe that the sensing resistors dissipate very little power.)

P_T	

Name ______________________

Date ______________________

Class ______________________

LAB 19

Series-Parallel ac Circuits

OBJECTIVES

After completing this lab, you will be able to

- analyze a series-parallel circuit to determine the current through and voltage across each element in a series-parallel circuit,
- measure voltage across each element in a series-parallel circuit using an oscilloscope and use the measurements to determine the current through each element of a series-parallel circuit,
- calculate the power dissipated by each element in a circuit,
- use measurements to verify that the actual powers dissipated correspond to theory.

EQUIPMENT REQUIRED

☐ Dual-trace oscilloscope
☐ Signal generator (sinusoidal function generator)
Note: Record this equipment in Table 19-1.

COMPONENTS

☐ Resistors: 10-Ω (2), 470-Ω (1/4-W carbon, 5% tolerance)
☐ Capacitors: 3300-pF (10% tolerance)
☐ Inductors: 1-mH (iron core, 5% tolerance)

EQUIPMENT USED

TABLE 19-1

Instrument	Manufacturer/Model No.	Serial No.
Oscilloscope		
Signal generator		

TEXT REFERENCE

Section 18.6 SERIES-PARALLEL CIRCUITS

DISCUSSION

The equivalent impedance of a series-parallel ac circuit is determined in a manner which is similar to that used in finding the equivalent resistance of a series-parallel circuit, with the exception that vector algebra is used in determining the total impedance at a given frequency. It is necessary to decide which elements or branches are in series and which are in parallel. The resultant impedance is the combination of the various connections. Once we have the total impedance, it is a simple matter to calculate the total current. By applying appropriate circuit theory, the current, voltage, and power of the various components of the circuit may then be found.

Refer to the circuit of Figure 19-1. Notice that resistor R and inductor L are in parallel. This parallel connection is then seen to be in series with capacitor C. The total impedance of the circuit is therefore determined as

$$Z_T = -jX_C + R \parallel jX_L \tag{19-1}$$

The power provided to the circuit by the voltage source is calculated as

$$P_T = EI\cos\theta = \frac{E^2}{Z_T}\cos\theta \tag{19-2}$$

In the preceding expression, E and I are the rms values of the sinusoidal voltage e and current i. Z_T is the magnitude of the circuit impedance and θ is the angle between the current phasor $\mathbf{I} = \mathbf{I}_C$ and the voltage phasor $\mathbf{E}$. (This is the same angle as the angle in the impedance vector $\mathbf{Z}_T$.)

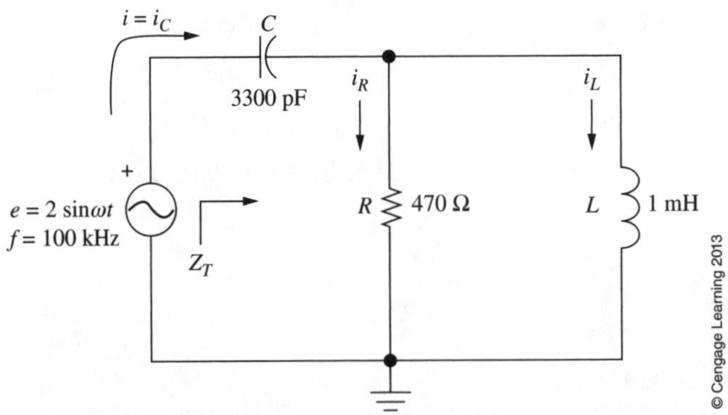

FIGURE 19-1 Series-parallel circuit.

On further examination of Figure 19-1, we see that only the resistor can dissipate power. (Inductors and capacitors do not dissipate power.) This means that the total power in the circuit must also be the same as the power dissipated by the resistor, namely

$$P_R = \frac{V_R^2}{R} = I_R^2 R \tag{19-3}$$

where V_R and I_R are rms quantities.

CALCULATIONS

1. Refer to the circuit of Figure 19-1. Determine the reactances of the capacitor and the inductor at a frequency of $f = 100$ kHz. Calculate the circuit impedance $\mathbf{Z}_T$. Enter all values in Table 19-2.

TABLE 19-2

X_C	
X_L	
$\mathbf{Z}_T$	

2. Convert the time-domain form of the voltage source into its equivalent phasor-domain form. Calculate the phasor current through each element of the circuit. Enter your results in Table 19-3.

TABLE 19-3

$\mathbf{E}$	
$\mathbf{I}_R$	
$\mathbf{I}_C$	
$\mathbf{I}_L$	

3. Calculate and record the total power provided to the circuit by the voltage source.

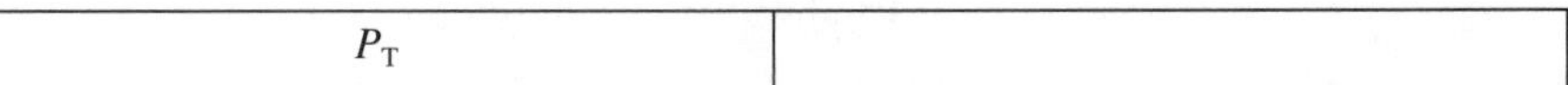

P_T	

4. Use complex algebra to show that currents $\mathbf{I}_R$, $\mathbf{I}_C$, and $\mathbf{I}_L$ satisfy Kirchhoff's current law

$$\sum \mathrm{I} = 0 \tag{19-4}$$

MEASUREMENTS

5. Assemble the circuit of Figure 19-2. Notice that two 10-Ω *sensing resistors* have been added to the circuit to help in determining branch currents. Since these resistors are small in comparison to the impedance in the branch, they will not significantly load the circuit.
6. Connect Ch1 of the oscilloscope to the output of the signal generator at point *a*. Set the oscilloscope to an automatic sweep and use Ch1 as the *trigger source*. Adjust the output of the generator to provide a sinusoidal voltage with an amplitude of 2.0 V (4.0 $V_{p\text{-}p}$) at a frequency of $f = 100$ kHz ($T = 10$ μs).
7. Use Ch2 of the oscilloscope to observe the voltage at point *c*. This is the voltage across sensing resistor R_1. Measure the peak-to-peak voltage V_1. Determine the phase angle θ_1 of voltage v_1 with respect to the generator voltage *e*. Write the time-domain expression for *i* using your measurements and calculations. Record your results in Table 19-4.

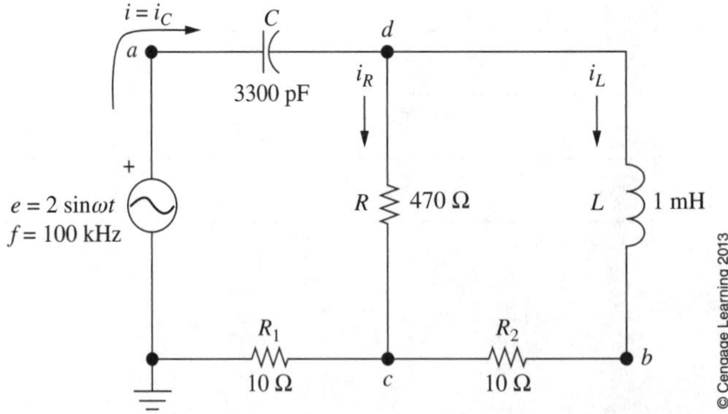

FIGURE 19-2 Using sensing resistors for current measurement.

TABLE 19-4

V_1	$V_{p\text{-}p}$
θ_1	
$i = i_c$	

8. Remove resistor R_1 from the circuit and replace it with a short circuit. Move Ch2 of the oscilloscope to observe the voltage at point d (the voltage across resistor R). Measure the peak-to-peak voltage V_R and determine phase angle θ_R of voltage v_R with respect to the generator voltage e. Record your results in Table 19-5. Determine the amplitude i_R and enter the time-domain expression for i_R in Table 19-5.

TABLE 19-5

V_R	$V_{p\text{-}p}$
θ_R	
i_R	

9. With Ch1 of the oscilloscope connected to point a, connect Ch2 of the oscilloscope to point b. Measure and record the peak-to-peak voltage across the sensing resistor R_2. Measure and record the phase angle between the generator voltage and the sensing voltage V_2. Use your measurements and Ohm's law to determine the sinusoidal expression for i_L. Record the result in Table 19-6.

TABLE 19-6

V_2	$V_{p\text{-}p}$
θ_2	
i_L	

CONCLUSIONS

10. Convert the measured currents i_R, i_C, and i_L of Table 19-4 through Table 19-6 into their phasor forms. Enter your results in Table 19-7 and compare them with the theoretical values of Table 19-3.

TABLE 19-7

$\mathbf{I} = \mathbf{I}_C$	
$\mathbf{I}_R$	
$\mathbf{I}_L$	

11. Use complex algebra to verify that the data of Table 19-7 satisfy Kirchhoff's current law.

12. Use the phasors **E** and **I** to calculate the total power delivered to the circuit by the voltage source. Enter your calculation in the space provided.

P_T	

13. Now use the rms values of V_R and I_R to determine the power dissipated by the resistor. Enter the result in the space below. How does this value compare with the total power delivered to the circuit by the voltage source? Offer an explanation for any variation.

P_R	

PROBLEMS

14. Refer to the circuit of Figure 19-1. If the frequency of the signal generator is increased to 200 kHz, determine the following:
    a. Total impedance $\mathbf{Z}_T$
    b. Currents $\mathbf{I}$, $\mathbf{I}_R$, and $\mathbf{I}_L$
    c. Power P_R dissipated by the resistor
    d. Power P_T delivered by the voltage source
15. Repeat Problem 14 if the frequency of the generator is decreased to 50 kHz.

Name ______________________

Date ______________________

Class ______________________

LAB 20

Thévenin's and Norton's Theorems (ac)

OBJECTIVES

After completing this lab, you will be able to

- calculate the Thévenin and Norton equivalents of an ac circuit,
- measure the Thévenin (open circuit) voltage and the Norton (short circuit) current of an ac circuit,
- calculate the Thévenin impedance of a circuit using the measured values of Thévenin voltage and Norton current,
- measure the load impedance which results in a maximum transfer of power to the load.

EQUIPMENT REQUIRED

☐ Dual-trace oscilloscope
☐ Signal generator (sinusoidal function generator)
☐ Digital multimeter (DMM)
Note: Record this equipment in Table 20-1.

COMPONENTS

☐ Resistors: 10-Ω, 1.5-kΩ (1/4-W carbon, 5% tolerance)
5-kΩ variable resistor
☐ Capacitors: 2200-pF (10% tolerance)
☐ Inductors: 2.4-mH (iron core, 5% tolerance)

EQUIPMENT USED

TABLE 20-1

Instrument	Manufacturer/Model No.	Serial No.
Oscilloscope		
Signal generator		
DMM		

TEXT REFERENCE

Section 20.3 THÉVENIN'S THEOREM—INDEPENDENT SOURCES
Section 20.4 NORTON'S THEOREM—INDEPENDENT SOURCES
Section 20.6 MAXIMUM POWER TRANSFER THEOREM

DISCUSSION

Thévenin's theorem allows us to convert any two-terminal linear bilateral circuit into an equivalent circuit consisting of a voltage source $\mathbf{E}_{Th}$ in series with an impedance $\mathbf{Z}_{Th}$ as illustrated in Figure 20-1(a). When a load is connected across the two terminals of the circuit, the current through (or voltage across) the load is easily calculated by analyzing the equivalent circuit. The Thévenin voltage of an equivalent circuit is determined by removing the load from the circuit and measuring the open-circuit voltage.

Norton's theorem is the complement of Thévenin's theorem in that it converts any two-terminal linear bilateral circuit into an equivalent circuit consisting of a current source $\mathbf{I}_N$ in parallel with an impedance $\mathbf{Z}_N$ as shown in Figure 20-1(b). The Norton current of an equivalent circuit is determined by replacing the load with a short circuit and measuring the current through the load.

Unlike dc circuits, the Thévenin (and Norton) impedance of an ac circuit cannot be measured directly. Rather, the impedance is determined indirectly by using the equivalence between the Thévenin and Norton circuits. Since the circuits are equivalent, the following relationship must apply:

$$\mathbf{Z}_{Th} = \mathbf{Z}_N = \frac{\mathbf{E}_{Th}}{\mathbf{I}_N} \qquad (20\text{-}1)$$

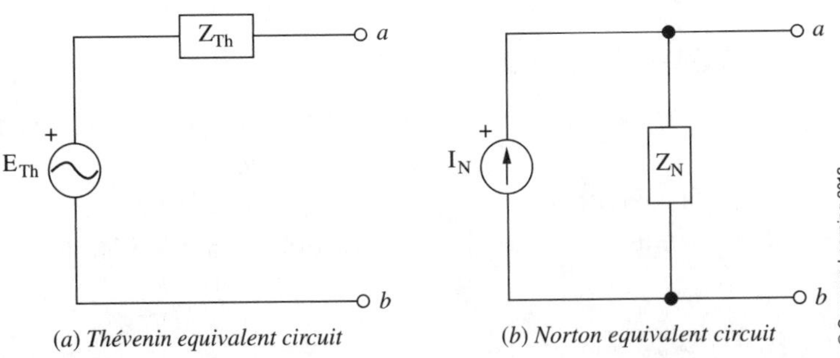

(a) Thévenin equivalent circuit   (b) Norton equivalent circuit

FIGURE 20-1

CALCULATIONS

1. Refer to the circuit of Figure 20-2. Determine the reactances of the capacitor and the inductor at a frequency of $f = 100$ kHz. Enter the results in Table 20-2.

TABLE 20-2

X_C	
X_L	

2. Convert the time-domain form of the voltage source in Figure 20-2 into its equivalent phasor-domain form and enter the result here.

E	

3. Determine the Thévenin equivalent circuit to the left of terminals a and b in the circuit of Figure 20-2. Sketch the equivalent circuit in this space.

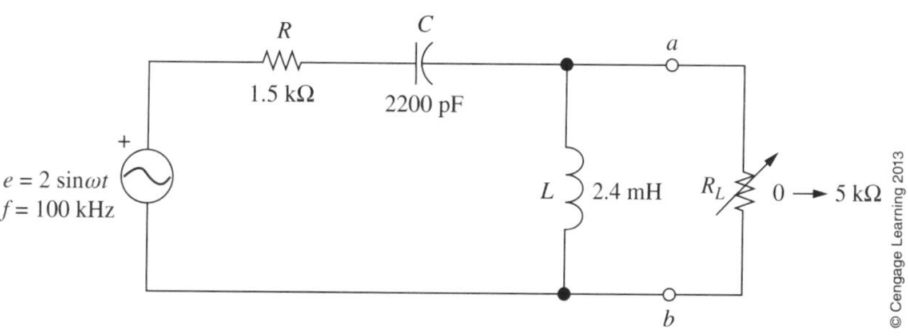

FIGURE 20-2

4. Determine the Norton equivalent circuit to the left of terminals *a* and *b* in the circuit of Figure 20-2. Sketch the equivalent circuit in this space.

5. *Absolute maximum power* will be delivered to the load impedance when the load is the complex conjugate of the Thévenin (or Norton) impedance. For what value of load impedance $\mathbf{Z}_L$ will the circuit of Figure 20-2 transfer maximum power to the load? Solve for the absolute maximum load power. Enter the results in Table 20-3.

TABLE 20-3

$\mathbf{Z}_L$	
P_{max}	

6. The load in the circuit of Figure 20-2 does not have any reactance. In this case, absolute maximum power cannot be transferred to the load. *Relative maximum power* will be transferred to the load impedance when the value of load resistance is equal to the following:

$$R = \sqrt{R_{Th}^2 + X_{Th}^2} \tag{20-2}$$

For what value of load resistance R_L will the circuit of Figure 20-2 transfer maximum power to the load? Solve for the relative maximum load power and enter the results in Table 20-4.

TABLE 20-4

R_L	
P_{max}	

MEASUREMENTS

7. Assemble the circuit of Figure 20-2, temporarily omitting the load resistor R_L. Connect Ch1 of the oscilloscope to the output of the signal generator and adjust the generator to provide an output of 2.0 V_p (4.0 $V_{p\text{-}p}$) at f = 100 kHz.
8. While using Ch1 as the trigger source of the oscilloscope, connect Ch2 between terminals *a* and *b* of the circuit. Measure and record the peak-to-peak value of the open-circuit voltage amplitude E_{Th} and the phase shift. (The phase shift is measured with respect to the generator voltage observed on Ch1.) Enter the results in Table 20-5.

TABLE 20-5

E_{Th}	$V_{p\text{-}p}$
P_{max}	

9. Place a 10-Ω *sensing resistor* between terminals *a* and *b*. The sensing resistor has a very low impedance with respect to the other components in the circuit and so has a minimal loading effect. Readjust the generator voltage to ensure that the output is 2.0 V_p. Measure the peak-to-peak voltage across the sensing resistor and determine the peak-to-peak value of the "short circuit" current I_N and the phase shift θ. Enter the results in Table 20-6.

TABLE 20-6

V_{ab}	$V_{p\text{-}p}$
I_N	$mA_{p\text{-}p}$
θ	

10. Use the DMM to adjust the 5-kΩ variable resistor for a resistance of 500 Ω. Remove the sensing resistor and insert the variable resistor between terminals *a* and *b* of the circuit. Adjust the supply voltage for 2.0 $V_{p\text{-}p}$ and measure the amplitude of the output voltage, V_L. Enter your measurement in Table 20-7.
11. Remove R_L from the circuit and incrementally increase the resistance by 500 Ω. Repeat Step 10. Keep increasing the load resistance until R_L = 5 kΩ. Enter all data in Table 20-7.

TABLE 20-7

R_L	V_L
500 Ω	
1000 Ω	
1500 Ω	
2000 Ω	
2500 Ω	
3000 Ω	
3500 Ω	
4000 Ω	
4500 Ω	
5000 Ω	

CONCLUSIONS

12. Convert the measured voltage and phase angle for the Thévenin voltage of Step 8 into its correct phasor form. Record your result in the space provided below. How does this value compare to the theoretical value determined in Step 3?

$\mathbf{E}_{Th}$	

13. Convert the measured current and phase angle for the Norton current of Step 9 into its correct phasor form. Enter the result below. How does this value compare to the theoretical value determined in Step 4?

$\mathbf{I}_N$	

14. Use the phasors of Steps 12 and 13 to calculate the Thévenin (and Norton) impedance. How does this value compare to the theoretical value determined in Steps 3 and 4?

$\mathbf{Z}_{Th} = \mathbf{Z}_N$	

____

____

15. Use the data of Table 20-7 to calculate the power delivered to the load for each of the resistor values. (Remember that you will need to convert each voltage measurement into its equivalent rms value in order to calculate the power.) Enter the results in Table 20-8.

TABLE 20-8

R_L	P_L
500 Ω	
1000 Ω	
1500 Ω	
2000 Ω	
2500 Ω	
3000 Ω	
3500 Ω	
4000 Ω	
4500 Ω	
5000 Ω	

16. Use the data of Table 20-7 to sketch a graph of power (in microwatts) versus load resistance (in ohms). Connect the points with the best smooth, continuous curve. (A correctly drawn curve will not be drawn from point-to-point. Alternatively, your instructor may have you use Excel to graph your data.)
17. Use the curve of Graph 20-1 to determine the approximate value of load resistance R_L for which the load receives maximum power from the circuit. Enter the result here. How does this value compare to the value determined in Step 6?

R_L	

____

____

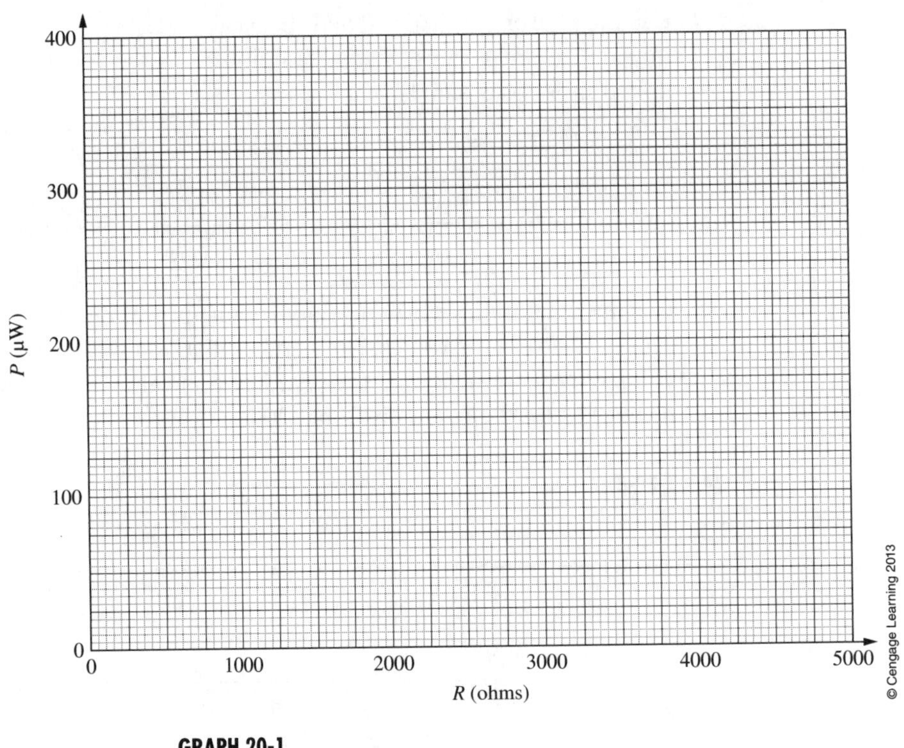

GRAPH 20-1

PROBLEMS

18. If the circuit of Figure 20-2 (f = 100 kHz) is to provide absolute maximum power to the load, what value of capacitance (in μF) or inductance (in mH) must be added in series with the load resistance?
19. Determine the Norton equivalent of the circuit to the left of points a and b in the circuit of Figure 20-2. Assume that the circuit operates at f = 200 kHz.
20. Refer to the circuit of Figure 20-2.
    a. Determine the Thévenin equivalent of the circuit to the left of points a and b, assuming that the circuit operates at a frequency of f = 50 kHz.
    b. Solve for the load impedance which will result in a maximum transfer of power to the load.
    c. Calculate the maximum power which can be transferred to the load.

Name ______________________

Date ______________________

Class ______________________

LAB 21

Series Resonance

OBJECTIVES

After completing this lab, you will be able to

- calculate the resonant frequency of a series resonant circuit,
- solve for the maximum output voltage of a resonant circuit using the *quality factor Q* of the circuit,
- measure the bandwidth of a series resonant circuit,
- measure the impedance at frequencies above and below the resonant frequency and observe that it is purely resistive only at resonance,
- sketch the circuit current as a function of frequency and explain why the response has a bell-shaped curve when plotted on a semilogarithmic graph.

EQUIPMENT REQUIRED

☐ Dual-trace oscilloscope
☐ Signal generator (sinusoidal function generator)
☐ Digital multimeter (DMM)
Note: Record this equipment in Table 21-1.

COMPONENTS

☐ Resistors: 15-Ω (1/4-W carbon, 5% tolerance)
☐ Capacitors: 0.33-μF (10% tolerance)
☐ Inductors: 2.4-mH (iron core, 5% tolerance)

EQUIPMENT USED

TABLE 21-1

Instrument	Manufacturer/Model No.	Serial No.
Oscilloscope		
Signal generator		
DMM		

TEXT REFERENCE

Section 21.1 SERIES RESONANCE
Section 21.2 QUALITY FACTOR, Q
Section 21.3 IMPEDANCE OF A SERIES RESONANT CIRCUIT
Section 21.4 POWER, BANDWIDTH, AND SELECTIVITY OF A SERIES RESONANT CIRCUIT

DISCUSSION

Resonant circuits are used throughout electronics as a means of passing a range of frequencies, while rejecting all other frequencies. These circuits have important applications in communications, where they are used in circuits such as receivers to tune into a particular station or channel. Figure 21-1 represents a typical series resonant circuit.

At the resonant frequency, the reactance of the inductor is exactly equal to the reactance of the capacitor. Since they are equal with opposite phase, the reactances cancel, and the total impedance of the circuit is purely resistive at resonance. The resonant frequency (in hertz) of a series circuit is given as

$$f_S = \frac{1}{2\pi\sqrt{LC}} \tag{21-1}$$

At the resonant frequency, the current (and power) in the circuit is maximum, resulting in a maximum output voltage appearing across the inductor. Since the reactance of the inductor can be many times greater than the resistance of the circuit, the output voltage may be many times greater than the applied signal. This characteristic is one of the advantages of using a resonant circuit since the output voltage is amplified without the need for *active components* such as transistors.

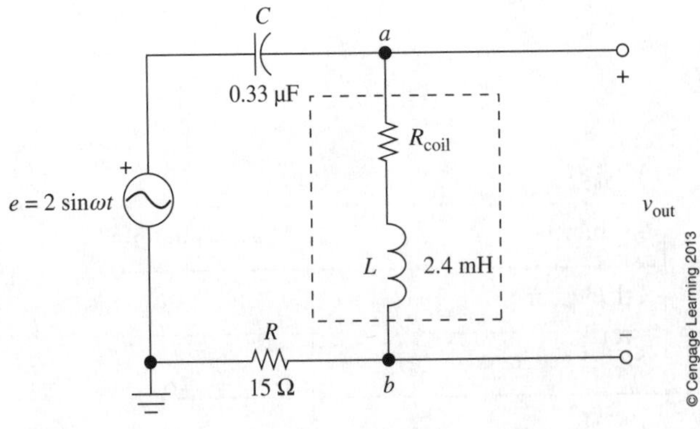

FIGURE 21-1 Series resonant circuit.

The *quality factor Q* of a resonant circuit is defined as the ratio of reactive power to the real power at the resonant frequency. It can be shown that the quality factor for the circuit of Figure 21-1 is determined as

$$Q = \frac{X_L}{R} = \frac{\omega L}{R_S + R_{\text{coil}}} \tag{21-2}$$

The Q of a circuit is used to determine the range of frequencies which will be passed by a given resonant circuit. If the circuit has a high Q (greater than 10), it will pass a narrow range of frequencies, and the circuit is said to have a *high selectivity*. Conversely, if the Q of the circuit is small, the circuit will pass a broader range of frequencies and the circuit is said to have a *low selectivity*. The *bandwidth (BW)* of a resonant circuit is defined as the difference between the half-power frequencies, namely the frequencies at which the circuit dissipates half the power that would be dissipated at resonance. The bandwidth (in hertz) of a resonant circuit is determined as

$$BW = \frac{f_S}{Q} \tag{21-3}$$

The half-power frequencies occur on either side of the resonant frequency. If the quality factor of the circuit is large ($Q \geq 10$) then the half-power frequencies are given approximately as

$$f_1 \cong f_S - \frac{BW}{2} \tag{21-4}$$

and

$$f_2 \cong f_S + \frac{BW}{2} \tag{21-5}$$

CALCULATIONS

1. Determine the resonant frequency for the circuit of Figure 21-1.

f_S	

2. Prior to starting the lab, obtain an inductor from your lab instructor. Use the DMM ohmmeter to measure the dc resistance of the inductor. Enter the result below.

R_{coil}	

3. Using the measured resistance of the inductor, calculate the phasor form of current **I** at resonance. Enter the result in the space provided below.

I	

4. Determine the phasor form of output voltage $\mathbf{V}_{out}$ appearing across the inductor. Enter the result below. You will need to consider the effect of R_{coil}.

$\mathbf{V}_{out}$	

5. Calculate the quality factor Q, bandwidth BW, and approximate half-power frequencies f_1 and f_2 for the circuit. Record the results in Table 21-2.

TABLE 21-2

Q	
BW	
f_1	
f_2	

MEASUREMENTS

6. Assemble the circuit shown in Figure 21-1. Connect Ch1 of the oscilloscope to the output of the signal generator and adjust the generator to provide an amplitude of 2.0 V_p (4.0 $V_{p\text{-}p}$) at a frequency of $f = 1$ kHz.
7. Connect Ch2 of the oscilloscope across the resistor R and measure the amplitude of V_R. (In order to simplify this measurement, it is generally easier to measure the peak-to-peak voltage and then divide by 2.) Use the measured voltage to calculate the amplitude of the current I. Enter the results in Table 21-3.
8. Increase the frequency of the signal generator to the frequencies indicated in Table 21-3. Due to loading effects, the amplitude of the generator will tend to drift. Ensure that the output of the signal generator is kept constant at 2.0 V_p. Measure the amplitude of V_R for each frequency and calculate the corresponding amplitude of current I. Enter the values in Table 21-3.

TABLE 21-3

f	V_R	I
1 kHz		
2 kHz		
3 kHz		
4 kHz		
4.5 kHz		
5 kHz		
5.5 kHz		
6 kHz		
6.5 kHz		
7 kHz		
8 kHz		
9 kHz		
10 kHz		

9. For which frequency f in Table 21-3 is the circuit current a maximum? Adjust the generator to provide an output of 2.0 V_p at this frequency. While observing the oscilloscope, adjust the generator frequency until the resistor voltage V_R is at the maximum value. Record the resonant frequency f_S and the corresponding resistor voltage V_R in Table 21-4. Calculate the current at resonance.
10. Decrease the frequency until the output voltage V_R is reduced to 0.707 of the maximum value found in Step 9. Record the lower half-power frequency f_1 and the corresponding resistor voltage V_R in Table 21-4. (Ensure that the output of the signal generator is at 2.0 V_p.)

    Increase the frequency above the resonant frequency until the output voltage V_R is again reduced to 0.707 of the maximum value found in Step 9. Record the upper half-power frequency f_2 and the corresponding resistor voltage V_R in Table 21-4. (Ensure that the output of the signal generator is at 2.0 V_p.)

    Calculate the current for each frequency.

TABLE 21-4

f	V_R	I
f_1 =		
f_S =		
f_2 =		

11. Set the signal generator to the resonant frequency determined in Step 9 and adjust the amplitude for 2.0 V_p. With Ch1 of the oscilloscope at the output of the signal generator, use Ch2 to measure the voltage across resistor *R*. Calculate the phasor form of current **I** at resonance. (You should observe that v_R and e are in phase.) Record your results in Table 21-5.

TABLE 21-5

V_R	
θ	
$\mathbf{I}$	

12. Adjust the signal generator for a frequency $f = f_1$ and a voltage of 2.0 V_p. Measure the magnitude and phase angle of the voltage V_R. Calculate the phasor form of current **I** at this frequency. Enter the results in Table 21-6.

TABLE 21-6

V_R	
θ_1	
$\mathbf{I}_1$	

13. Adjust the signal generator for a frequency $f = f_2$ and a voltage of 2.0 V_p. Measure the magnitude and phase angle of the voltage V_R. Calculate the phasor form of the current **I** at this frequency. Enter the results in Table 21-7.

TABLE 21-7

V_R	
θ_2	
$\mathbf{I}_2$	

14. Set the signal generator to the resonant frequency determined in Step 9 and adjust the amplitude for 2.0 V_p. Place Ch1 of the oscilloscope at terminal *a* of the circuit and Ch2 at terminal *b*. Use the difference mode of the oscilloscope to measure the amplitude of the output voltage. You should observe that the amplitude of this voltage is larger than the amplitude at the output of the signal generator. Record the amplitude of output voltage V_{out} in the space provided.

V_{out}	

CONCLUSIONS

15. Plot the data of Tables 21-3 and 21-4 on the semilogarithmic scale of Graph 21-1. Connect the points with the best smooth, continuous curve. (Alternatively, your instructor may have you use Excel to graph your data.)

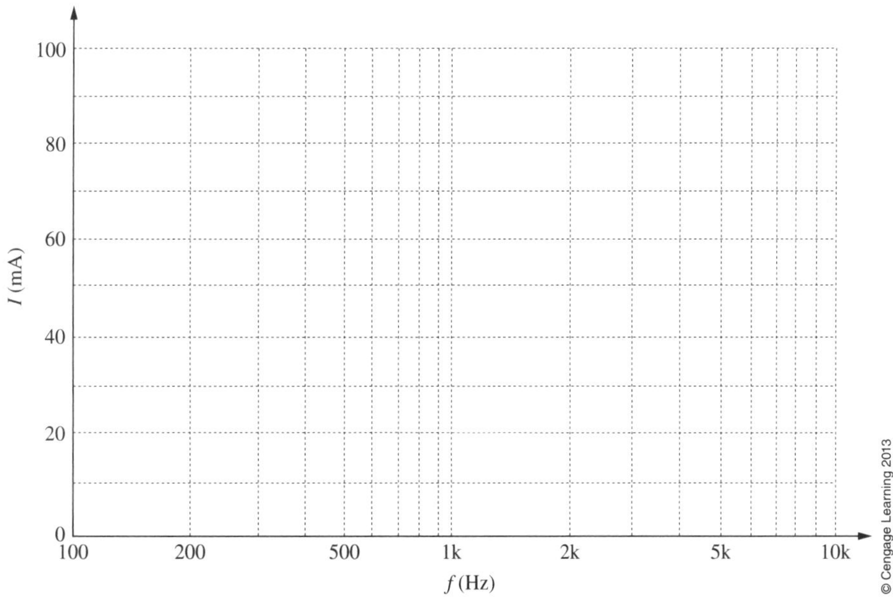

GRAPH 21-1

16. Compare the measured resonant frequency f_S found in Step 9 to the theoretical frequency determined in Step 1.

____

____

17. Compare the measured half-power frequencies from Step 10 to the theoretical values determined in Step 5.

____

____

____

____

18. Use the measured half-power frequencies to calculate the bandwidth of the circuit. Enter the results below.

$$BW = f_2 - f_1 \tag{21-6}$$

BW	

19. Calculate the Q of the circuit and enter the result in the space provided.

$$Q = \frac{f_S}{BW} \tag{21-7}$$

Q	

20. Use the data of Step 11 to compare the phase angle of the current with respect to the signal generator at resonance. Based on this result, is the circuit resistive, inductive, or capacitive when $f = f_S$?

____

____

21. Use the data of Step 12 to compare the phase angle of the current with respect to the signal generator when $f = f_1$. Based on this result, is the circuit resistive, inductive, or capacitive when $f < f_S$?

____

____

22. Use the data of Step 13 to compare the phase angle of the current with respect to the signal generator when $f = f_2$. Based on this result, is the circuit resistive, inductive, or capacitive when $f > f_S$?

____

____

23. In Step 14, you should have observed that the output voltage of the circuit at resonance is larger than the applied signal generator voltage. Calculate and record the ratio of the amplitudes V_{out}/E. Since the output voltage is taken across the inductor, you should observe that this ratio is very close to the expected Q of the circuit. Compare this result with that obtained in Step 19.

$Q = V_{out}/E$	

___

___

FOR FURTHER INVESTIGATION AND DISCUSSION

24. If the resistance of the inductor R_{coil} was higher than the measured value, what would happen to f_S, Q, BW, and the output voltage v_{out} at resonance?

f_S:

___

___

Q:

___

___

BW:

___

___

v_{out} at resonance:

___

___

25. Use Multisim or PSpice to simulate the circuit of Figure 21-1. Use a coil resistance of $R_{coil} = 5\ \Omega$. Find the resonant frequency, bandwidth, and the quality factor of the circuit.
26. Repeat Step 25 by letting $R_{coil} = 10\ \Omega$.

Name ______________________

Date ______________________

Class ______________________

LAB 22

Parallel Resonance

OBJECTIVES

After completing this lab, you will be able to

- calculate the resonant frequency of a parallel resonant circuit,
- solve for the maximum output voltage of a parallel resonant circuit,
- measure the bandwidth of a parallel resonant circuit,
- measure the impedance at frequencies above and below the resonant frequency and observe that it is purely resistive at resonance,
- sketch the output voltage as a function of frequency and explain why the response has a bell-shaped curve when plotted on a semilogarithmic graph.

EQUIPMENT REQUIRED

☐ Dual-trace oscilloscope
☐ Signal generator (sinusoidal function generator)
☐ Digital multimeter (DMM)
Note: Record this equipment in Table 22-1.

COMPONENTS

☐ Resistors: 1-kΩ, 1.5-kΩ (1/4-W carbon, 5% tolerance)
☐ Capacitors: 0.33-μF (10% tolerance)
☐ Inductors: 2.4-mH (iron core, 5% tolerance)

EQUIPMENT USED

TABLE 22-1

Instrument	Manufacturer/Model No.	Serial No.
Oscilloscope		
Signal generator		
DMM		

TEXT REFERENCE

Section 21.2 QUALITY FACTOR, Q
Section 21.6 PARALLEL RESONANCE

DISCUSSION

Although series resonant networks are used occasionally in electrical and electronic circuits, parallel resonant networks are the most common type used. Figure 22-1 illustrates a simple parallel resonant network, often called an *LC tank circuit*.

The impedance of the tank circuit is relatively low at all frequencies except at the frequency of resonance. At the resonant frequency, the reactance of the capacitor is exactly equal to the reactance of the inductor. The resulting parallel impedance approaches that of an open circuit. Since the inductor will always have some series resistance due to the coil of wire, the actual impedance of the tank circuit will not be infinitely large. In practice, the impedance of the tank circuit will generally have an impedance between 10 kΩ and 100 kΩ. When the tank circuit is connected across a constant current source (usually a transistor), the voltage across the tank circuit will be relatively high at the resonant frequency and very low at all other frequencies.

The resonant frequency of a tank circuit is found to be

$$f_P = \frac{1}{2\pi\sqrt{LC}}\sqrt{1 - \frac{R_{coil}^2 C}{L}} \qquad (22\text{-}1)$$

If R_{coil}^2 is at least 10 times smaller than the ratio L/C, then the parallel resonant frequency may be approximated as

$$f_P = \frac{1}{2\pi\sqrt{LC}} \qquad (22\text{-}2)$$

The input impedance of a tank circuit at resonance will always be purely resistive, and may be determined by using the *quality factor Q* of the coil as follows:

$$R_P = (Q_{coil}^2 + 1)R_{coil} \qquad (22\text{-}3)$$

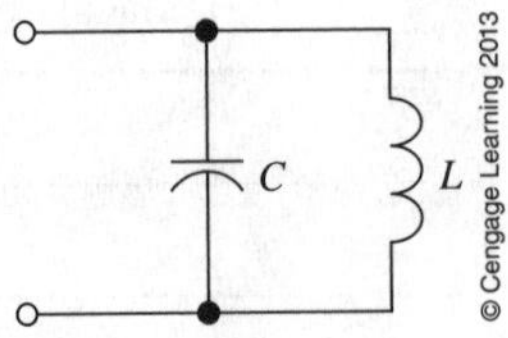

FIGURE 22-1 Ideal LC tank circuit.

If a resistance R_1 is placed in parallel with the tank, the Q of the circuit will be reduced since this resistor absorbs some of the energy from the circuit. Also, if the voltage source has a series resistance R_s, then the Q of the circuit is reduced still further. For the network shown, the quality factor is determined as

$$Q = \frac{R_{eq}}{X_c} \quad \text{where } R_{eq} = R_1 \,||\, R_P \,||\, R_S \tag{22-4}$$

Notice that the Q of the circuit is determined by placing R_s in parallel with the other resistors. The reason for this becomes apparent if the voltage source and its series resistance are converted into an equivalent current source and parallel resistance.

As in the series resonant circuit, the quality factor may be used to determine the bandwidth of the circuit as

$$BW = \frac{f_P}{Q} \tag{22-5}$$

CALCULATIONS

1. Calculate and record the resonant frequency for the circuit of Figure 22-2.

f_P	

2. Prior to starting the lab, obtain an inductor from your lab instructor. Use the DMM ohmmeter to measure the dc resistance of the inductor. Enter the result below.

R_{coil}	

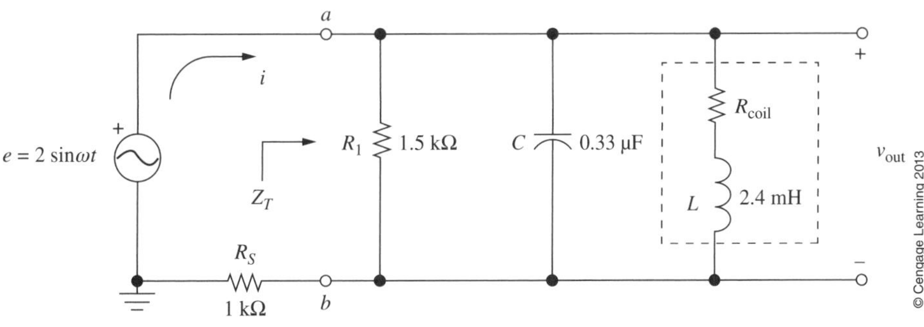

FIGURE 22-2 Parallel resonant circuit.

3. Calculate the equivalent impedance R_P of the LC tank at resonance. Use this value to determine the total impedance $\mathbf{Z}_T$ of the circuit at resonance. (The impedance will be resistive.) Calculate the phasor form of current $\mathbf{I}$ at resonance and determine the output voltage phasor $\mathbf{V}_{out}$. Enter your results in Table 22-2.

TABLE 22-2

R_P	
$\mathbf{Z}_T$	
$\mathbf{I}$	
$\mathbf{V}_{out}$	

4. Calculate and record the quality factor Q, bandwidth BW, and half-power frequencies f_1 and f_2 for the circuit. Enter your results in Table 22-3.

TABLE 22-3

Q	
BW	
f_1	
f_2	

MEASUREMENTS

5. Assemble the circuit shown in Figure 22-2. Connect Ch1 of the oscilloscope to the output of the signal generator and adjust the output to have an amplitude of 2.0 V_p (4.0 $V_{p\text{-}p}$) at a frequency of $f = 1$ kHz.
6. Use Ch1 as the trigger source and connect Ch2 of the oscilloscope across the resistor R_S. Set the oscilloscope on the difference mode to display the amplitude of the sinusoidal output voltage. Measure the amplitude of voltage V_{out} and record the results in Table 22-4.
7. Increase the frequency of the signal generator to the frequencies indicated in Table 22-4. Adjust the amplitude of the signal generator to maintain an amplitude of 2.0 V_p for each frequency. Measure and record the amplitude of the output voltage for each frequency.

TABLE 22-4

f	V_{out}
1 kHz	
2 kHz	
3 kHz	
4 kHz	
4.5 kHz	
5 kHz	
5.5 kHz	
6 kHz	
6.5 kHz	
7 kHz	
8 kHz	
9 kHz	
10 kHz	

8. For which frequency f in Table 22-4 is the output voltage a maximum? Adjust the generator to provide an output of 2.0 V_p at this frequency. While observing the oscilloscope, adjust the generator frequency until voltage V_{out} is at the maximum value. Record the resonant frequency f_P and the corresponding output voltage V_{out} in Table 22-5.
9. Decrease the frequency until the output voltage V_{out} is reduced to 0.707 of the maximum value found in Step 8. Record the lower half-power frequency f_1 and the corresponding output voltage V_{out} in Table 22-5. (Ensure that the output of the signal generator is at 2.0 V_p.)

   Increase the frequency above the resonant frequency until the output voltage V_{out} is again reduced to 0.707 of the maximum value found in Step 8. Record the upper half-power frequency f_2 and the corresponding resistor voltage V_{out} in Table 22-5. (Ensure that the output of the signal generator is at 2.0 V_p.)

TABLE 22-5

f	V_{out}
$f_1 =$	
$f_P =$	
$f_2 =$	

10. Set the signal generator to the resonant frequency determined in Step 9 and adjust the amplitude for 2.0 V_p. Measure the magnitude and phase angle (with respect to the signal generator) of the voltage across R_S. Calculate the phasor form of current I at resonance. (You should observe that v_S and e are in phase.) Record your results in Table 22-6.

TABLE 22-6

V_S	
θ	
I	

11. Adjust the signal generator for a frequency $f = f_1$ and a voltage of 2.0 V_p. Measure the magnitude and phase angle of the voltage V_S. Calculate the phasor form of current **I** at this frequency. Record these values in Table 22-7.

TABLE 22-7

V_R	
θ	
I	

12. Adjust the signal generator for a frequency $f = f_2$ and a voltage of 2.0 V_p. Measure the magnitude and phase angle of the voltage V_S. Calculate the phasor form of the current **I** at this frequency. Record the values in Table 22-8.

TABLE 22-8

V_S	
θ	
I	

CONCLUSIONS

13. Plot the data of Tables 22-4 and 22-5 on the semilogarithmic scale of Graph 22-1. Connect the points with the best smooth, continuous curve. (Alternatively, your instructor may have you use Excel to graph your data.)

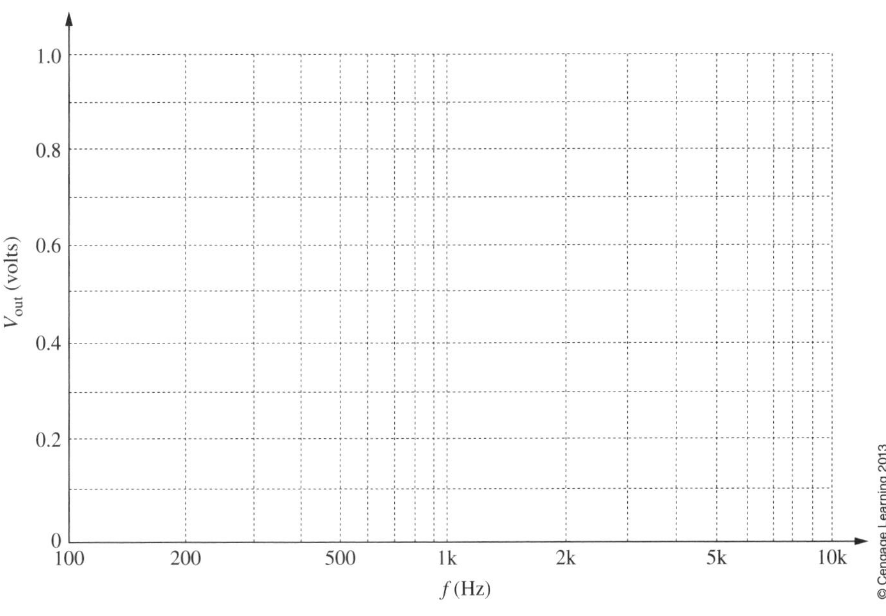

GRAPH 22-1

14. Compare the measured resonant frequency f_P as found in Step 8 to the theoretical frequency determined in Step 1.

____

____

15. Compare the measured half-power frequencies of Step 9 to the theoretical values determined in Step 3.

____

____

16. Calculate and record the bandwidth.

$$BW = f_2 - f_1 \tag{22-6}$$

BW	

17. Calculate and record the quality factor Q of the circuit.

$$Q = \frac{f_S}{BW} \tag{22-7}$$

Q	

Compare this value of Q to that calculated in Step 3.

18. Use the data of Step 10 to compare the phase angle of the current with respect to the signal generator at resonance. Based on this result, is the circuit resistive, inductive, or capacitive when $f = f_P$?

19. Use the data of Step 11 to compare the phase angle of the current with respect to the signal generator when $f = f_1$. Based on this result, is the circuit resistive, inductive, or capacitive when $f < f_P$?

20. Use the data of Step 12 to compare the phase angle of the current with respect to the signal generator when $f = f_2$. Based on this result, is the circuit resistive, inductive, or capacitive when $f > f_P$?

FOR FURTHER INVESTIGATION AND DISCUSSION

21. If the resistance of the inductor R_{coil} were higher than the measured value, what would happen to R_P, Q, and BW at resonance?

R_P:

Q:

BW:

Name ______________________

Date ______________________

Class ______________________

LAB 23

RC and *RL* Low-Pass Filter Circuits

OBJECTIVES

After completing this lab, you will be able to

- develop the transfer function for a low-pass filter circuit,
- determine the cutoff frequency of a low-pass filter circuit,
- sketch the Bode plot of the transfer function for a low-pass filter,
- compare the measured voltage gain response of a low-pass filter to the theoretical asymptotic response predicted by a Bode plot,
- explain why the voltage gain of a low-pass filter drops at a rate of 20 dB for each decade increase in frequency.

EQUIPMENT REQUIRED

☐ Dual-trace oscilloscope
☐ Signal generator (sinusoidal function generator)
☐ Digital multimeter (DMM)
Note: Record this equipment in Table 23-1.

COMPONENTS

☐ Resistors:	75-Ω, 330-Ω (1/4-W carbon, 5% tolerance)
☐ Capacitors:	0.47-μF (10% tolerance)
☐ Inductors:	2.4-mH (iron core, 5% tolerance)

EQUIPMENT USED

TABLE 23-1

Instrument	Manufacturer/Model No.	Serial No.
Oscilloscope		
Signal generator		
DMM		

TEXT REFERENCE

Section 22.3 SIMPLE *RC* AND *RL* TRANSFER FUNCTIONS
Section 22.4 THE LOW-PASS FILTER

DISCUSSION

Filter circuits are used extensively in electrical and electronic circuits to remove unwanted signals while permitting desired signals to pass from one stage to another. Although there are many types of filter circuits, most filters are *low-pass, high-pass, band-pass,* or *band reject* filters. As the name implies, *low-pass* filters permit low frequencies to pass from one stage to another. Figure 23-1 shows both *RC* and *RL* low-pass filters.

The frequency at which the low-pass filter begins to attenuate (decrease the amplitude of) a signal is called the *cutoff frequency* or *break frequency* f_C. Specifically, the cutoff frequency is that frequency at which the amplitude of the output voltage is 0.707 of the amplitude of low-frequency signals. Since this frequency corresponds to half power, the cutoff frequency is also called the *half-power* or *3-dB down frequency*. (Recall that when the output power is half of the input power, the attenuation of the stage is 3 dB.)

Cutoff frequencies are generally calculated as ω_C in radians per second, since the algebra tends to be fairly straightforward. However, when working with measurements it is easiest to use frequencies expressed as f_C in hertz. The cutoff frequencies for an *RC* low-pass filter are given as follows:

$$\omega_C = \frac{1}{\tau} = \frac{1}{RC} \tag{23-1}$$

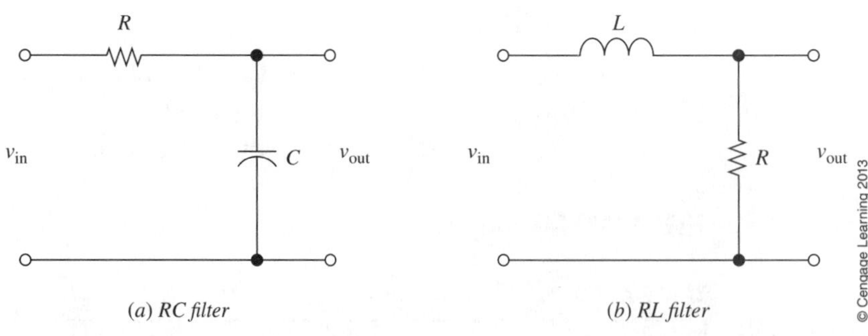

FIGURE 23-1 Low-pass filters.

and

$$f_C = \frac{\omega_C}{2\pi} = \frac{1}{2\pi RC} \tag{23-2}$$

For an *RL* low-pass filter, the cutoff frequencies are

$$\omega_C = \frac{1}{\tau} = \frac{R}{L} \tag{23-3}$$

and

$$f_C = \frac{\omega_C}{2\pi} = \frac{R}{2\pi L} \tag{23-4}$$

The frequency response of a filter circuit is generally shown on two semi-logarithmic graphs where the abscissa (horizontal axis) for each graph gives the frequency on a logarithmic scale. The ordinate (vertical axis) of one graph shows the voltage gain in decibels, while the ordinate of the other graph shows the phase shift of the output voltage with respect to the applied input voltage. The voltage gain and phase shift for any frequency are determined by finding the *transfer function TF* of the given filter. The transfer function is defined as the ratio of the output voltage phasor to the input voltage phasor.

$$\mathbf{TF} = \frac{V_{out}}{V_{in}} \tag{23-5}$$

CALCULATIONS

The *RC* Low-Pass Filter

1. Write the transfer function $\mathbf{TF} = \mathbf{V}_{out}/\mathbf{E}$ for the *RC* low-pass filter of Figure 23-2.

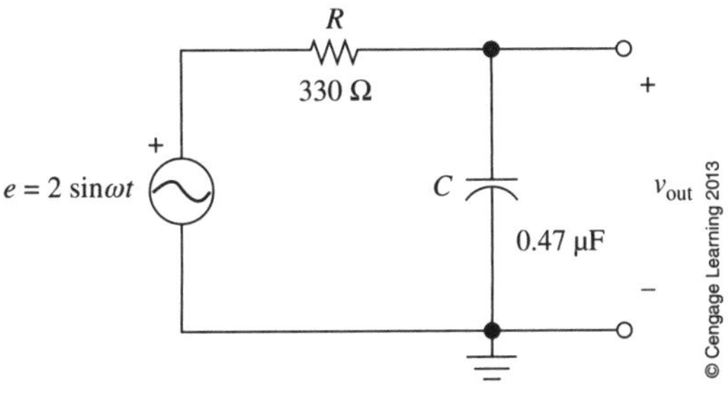

FIGURE 23-2 *RC* low-pass filter.

2. Determine the cutoff frequencies (in radians per second and in hertz) of the circuit in Figure 23-2. Record the values in Table 23-2.

TABLE 23-2

ω_C	rad/s
f_C	Hz

3. Sketch the straight-line approximations of the frequency responses (A_v in dB versus frequency and θ versus frequency) for the *RC* low-pass filter on Graph 23-1.

The *RL* Low-Pass Filter

4. Write the transfer function $\mathbf{TF} = \mathbf{V}_{out}/\mathbf{E}$ for the *RL* low-pass filter of Figure 23-3.

5. Determine the cutoff frequencies (in radians per second and in hertz) of the circuit in Figure 23-3. Record the values in Table 23-3.

TABLE 23-3

ω_C	rad/s
f_C	Hz

6. Sketch the straight-line approximations of the frequency responses (A_v in dB versus frequency and θ versus frequency) for the *RL* low-pass filter on Graph 23-2.

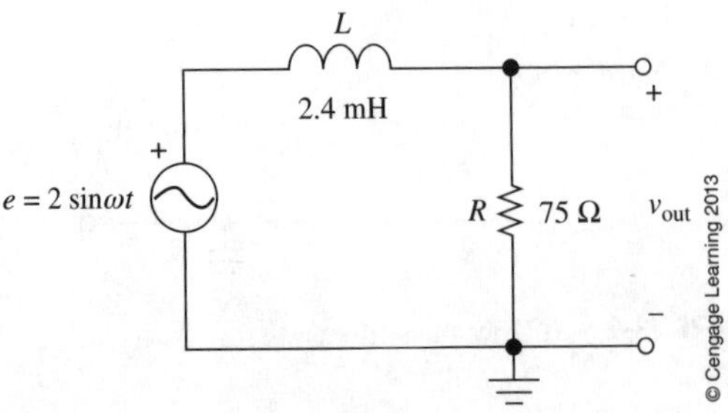

FIGURE 23-3 *RL* low-pass filter.

MEASUREMENTS

The *RC* Low-Pass Filter

7. Assemble the circuit shown in Figure 23-2. Connect Ch1 of the oscilloscope to the output of the signal generator and adjust the output to have an amplitude of 2.0 V_p (4.0 V_{p-p}) at a frequency of f = 100 Hz.
8. Use Ch1 as the trigger source and connect Ch2 of the oscilloscope across the capacitor *C*. Measure the amplitude and phase angle (with respect to *e*) of the sinusoidal output voltage v_{out}. Enter the results in Table 23-4.
9. Increase the frequency of the signal generator to the frequencies indicated in Table 23-4. If necessary, adjust the amplitude of the signal generator to ensure that the output is maintained at 2.0 V_p (4.0 V_{p-p}). Measure the amplitude and phase shift of v_{out} (with respect to *e*) for each frequency.

TABLE 23-4

f	v_{out}	
	Amplitude	Phase shift
100 Hz		
200 Hz		
400 Hz		
800 Hz		
1 kHz		
2 kHz		
4 kHz		
8 kHz		
10 kHz		

10. Determine the cutoff frequency by adjusting the frequency of the generator until the output voltage has an amplitude of V_{out} = (0.707)(2.0 V_p) = 1.41 V_p. Record the measured cutoff frequency below.

f_C	

The *RL* Low-Pass Filter

11. Construct the circuit shown in Figure 23-3. Connect Ch1 of the oscilloscope to the output of the signal generator and adjust the output to have an amplitude of 2.0 V_p (4.0 $V_{p\text{-}p}$) at a frequency of $f = 100$ Hz.
12. Use Ch1 as the trigger source and connect Ch2 of the oscilloscope across the resistor R. Measure the amplitude and phase angle (with respect to e) of the sinusoidal output voltage v_{out}. Enter the results in Table 23-5.
13. Increase the frequency of the signal generator to the frequencies indicated in Table 23-5. If necessary, adjust the amplitude of the signal generator to ensure that the output is maintained at 2.0 V_p (4.0 $V_{p\text{-}p}$). Measure the amplitude and phase shift of v_{out} (with respect to e) for each frequency.

TABLE 23-5

f	v_{out}	
	Amplitude	Phase shift
100 Hz		
200 Hz		
400 Hz		
800 Hz		
1 kHz		
2 kHz		
4 kHz		
8 kHz		
10 kHz		

14. Determine the cutoff frequency by adjusting the frequency of the generator until the output voltage has an amplitude of $V_{out} = (0.707)(2.0\ V_p) = 1.41\ V_p$. Record the measured cutoff frequency here.

f_C	

CONCLUSIONS

The *RC* Low-Pass Filter

15. Use your measurements recorded in Table 23-4 to calculate the magnitude of the gain as a ratio of the amplitudes V_{out}/V_{in}. Calculate the voltage gain in decibels as

$$[A_v]_{dB} = 20 \log \frac{V_{out}}{V_{in}} \qquad (23\text{-}6)$$

Record the calculated voltage gain and measured phase shift for each frequency in Table 23-6.

TABLE 23-6

f	$A_v = V_{out}/V_{in}$	$[A_v]_{dB}$	θ
100 Hz			
200 Hz			
400 Hz			
800 Hz			
1 kHz			
2 kHz			
4 kHz			
8 kHz			
10 kHz			

16. Plot the data of Table 23-6 on the semilogarithmic scales of Graph 23-1 (page 181). Connect the points with the best smooth, continuous curve. (Alternatively, your instructor may have you use Excel to graph your data.) You should find that the actual response is closely predicted by the straight-line approximations of the Bode plot.
17. Compare the measured cutoff frequency f_C of Step 10 to the theoretical value predicted by the transfer function in Step 2.

___

___

The *RL* Low-Pass Filter

18. Use your measurements recorded in Table 23-5 to calculate the magnitude of the gain as a ratio of the amplitudes V_{out}/V_{in}. Determine the voltage gain in decibels. Record the calculated voltage gain and measured phase shift for each frequency in Table 23-7.

TABLE 23-7

f	$A_v = V_{out}/V_{in}$	$[A_v]_{dB}$	θ
100 Hz			
200 Hz			
400 Hz			
800 Hz			
1 kHz			
2 kHz			
4 kHz			
8 kHz			
10 kHz			

19. Plot the data of Table 23-7 on the semilogarithmic scales of Graph 23-2 (page 181). Connect the points with the best smooth, continuous curve. (Alternatively, your instructor may have you use Excel to graph your data.) You should find that the actual response is closely predicted by the straight-line approximations of the Bode plot.
20. Compare the measured cutoff frequency f_C of Step 14 to the theoretical value predicted by the transfer function in Step 5.

____

____

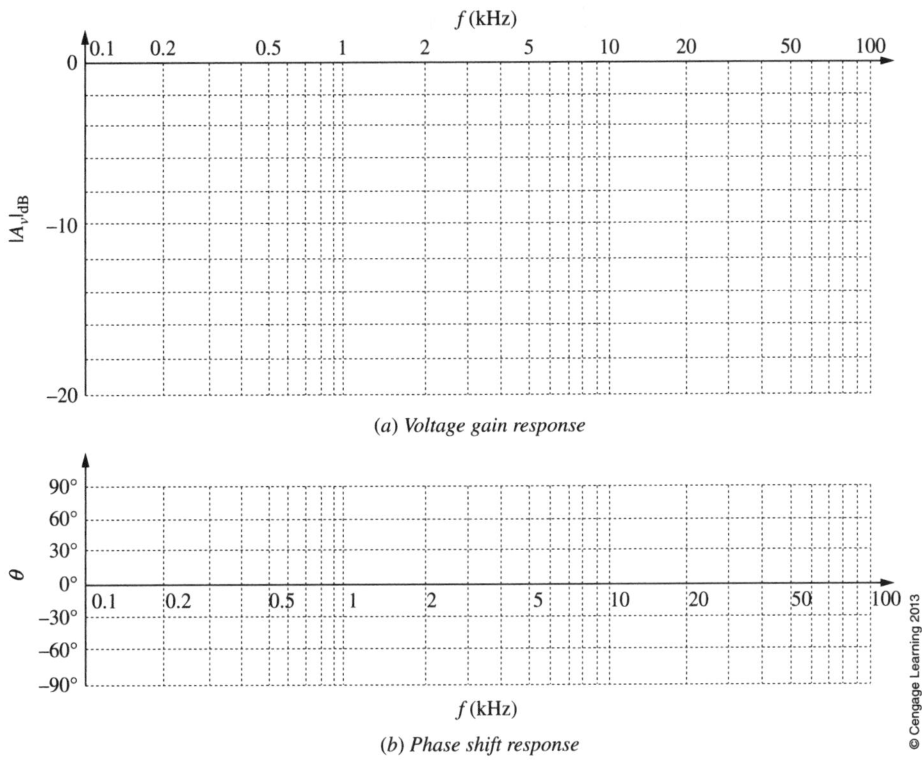

GRAPH 23-1 Frequency response of an *RC* low-pass filter.

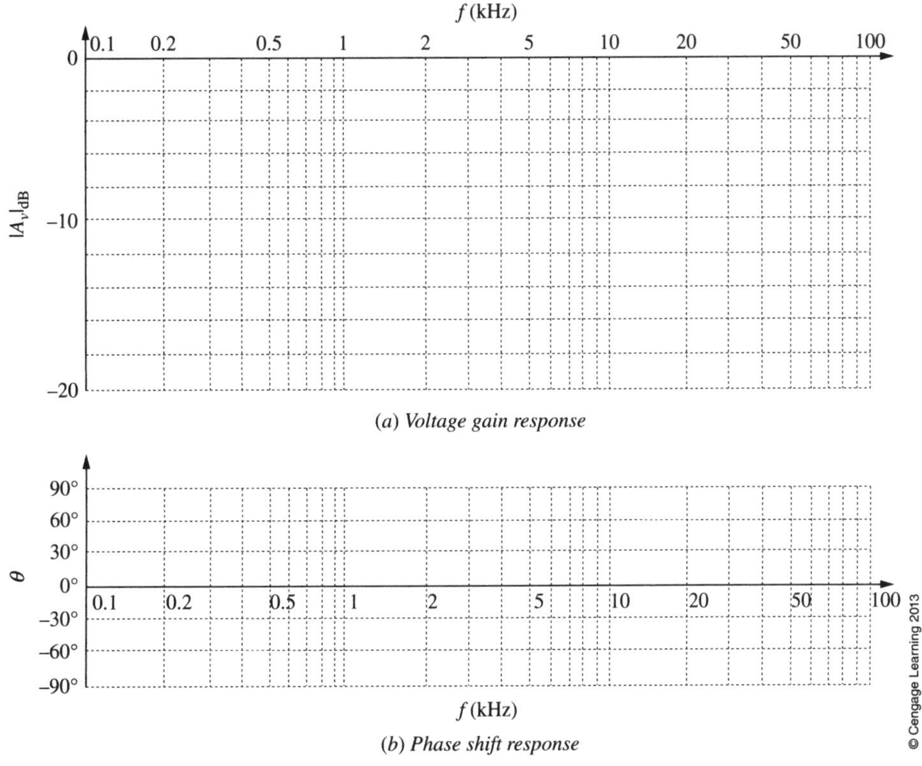

GRAPH 23-2 Frequency response of an *RL* low-pass filter.

FOR FURTHER INVESTIGATION AND DISCUSSION

21. Use Multisim or PSpice to simulate the circuit of Figure 23-2. Obtain the frequency response curves for this circuit and compare the results to the actual measurements.
22. Repeat Step 21 for the circuit of Figure 23-3.

Name ______________________

Date ______________________

Class ______________________

LAB 24

RC and *RL* High-Pass Filter Circuits

OBJECTIVES

After completing this lab, you will be able to

- develop the transfer function for a high-pass filter circuit,
- determine the cutoff frequency of a high-pass filter circuit,
- sketch the Bode plot of the transfer function for a high-pass filter,
- compare the measured voltage gain response of a high-pass filter to the theoretical asymptotic response predicted by a Bode plot,
- explain why the voltage gain of a high-pass filter increases at a rate of 20 dB/decade below the cutoff frequency.

EQUIPMENT REQUIRED

- ☐ Dual-trace oscilloscope
- ☐ Signal generator (sinusoidal function generator)
- ☐ Digital multimeters (DMM)

  Note: Record this equipment in Table 24-1.

COMPONENTS

☐ Resistors:	75-Ω, 330-Ω (1/4-W carbon, 5% tolerance)
☐ Capacitors:	0.47-μF (10% tolerance)
☐ Inductors:	2.4-mH (iron core, 5% tolerance)

EQUIPMENT USED

TABLE 24-1

Instrument	Manufacturer/Model No.	Serial No.
Oscilloscope		
Signal generator		
DMM		

TEXT REFERENCE

Section 22.5 THE HIGH-PASS FILTER

DISCUSSION

As the name implies, the *high-pass filter* permits high frequencies to pass from the input through to the output of the filter. Due to the abundance of electric motors and fluorescent lights, *60-Hz noise* is the most prevalent unwanted signal around us. Although many applications exist, one of the most common uses for the high-pass filter is to prevent 60-Hz noise from entering a sensitive electrical or electronic system. Figure 24-1 shows both *RC* and *RL* high-pass filters.

As in the low-pass filter circuit, the *cutoff frequency* f_C is that frequency at which the amplitude of the output voltage is 0.707 of the maximum amplitude. However, in the case of high-pass filters, the maximum amplitude occurs for high frequencies rather than for low frequencies. Since this frequency corresponds to half power, the cutoff frequency is also called the *half-power* or *3-dB down frequency*.

The cutoff frequencies for an *RC* high-pass filter are identical to the values for the low-pass *RC* filter and are given as follows:

$$\omega_C = \frac{1}{\tau} = \frac{1}{RC} \tag{24-1}$$

and

$$f_C = \frac{\omega_C}{2\pi} = \frac{1}{2\pi RC} \tag{24-2}$$

Similarly, for an *RL* high-pass filter, the cutoff frequencies are

$$\omega_C = \frac{1}{\tau} = \frac{R}{L} \tag{24-3}$$

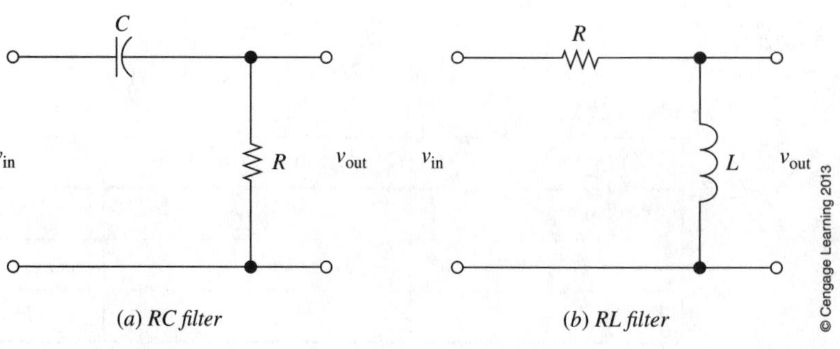

FIGURE 24-1 High-pass filters.

and

$$f_C = \frac{\omega_C}{2\pi} = \frac{R}{2\pi L} \qquad (24\text{-}4)$$

As in the low-pass filter, the frequency response of a high-pass filter circuit is generally shown on two semilogarithmic graphs where the abscissa (horizontal axis) for each graph gives the frequency on a logarithmic scale. The ordinate (vertical axis) of one graph shows the voltage gain in decibels, while the ordinate of the other graph shows the phase shift of the output voltage with respect to the applied input voltage. The voltage gain and phase shift for any frequency are determined by finding the *transfer function TF* of the given filter. Recall that the transfer function is defined as the ratio of the output voltage phasor to the input voltage phasor.

$$\text{TF} = \frac{V_{\text{out}}}{V_{\text{in}}} \qquad (24\text{-}5)$$

CALCULATIONS

The *RC* High-Pass Filter

1. Write the transfer function **TF** = $\mathbf{V}_{\text{out}}/\mathbf{E}$ for the *RC* high-pass filter of Figure 24-2.

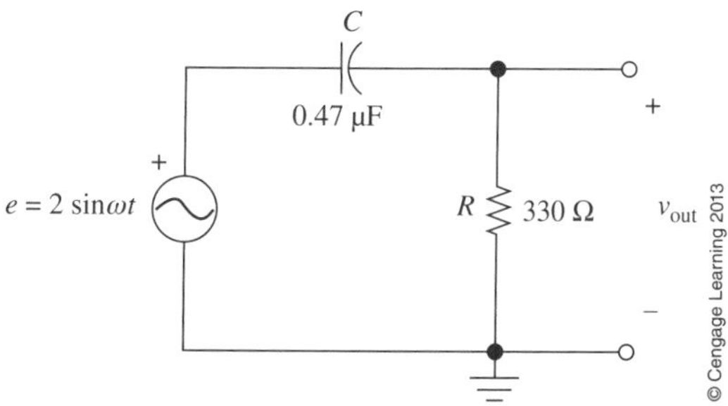

FIGURE 24-2 *RC* high-pass filter.

2. Determine the cutoff frequencies (in radians per second and in hertz) of the circuit in Figure 24-2. Record the values in Table 24-2.

TABLE 24-2

ω_C	rad/s
f_C	Hz

3. Sketch the straight-line approximations of the frequency responses (A_v in dB versus frequency and θ versus frequency) for the *RC* high-pass filter on Graph 24-1 (page 190).

The *RL* High-Pass Filter

4. Write the transfer function **TF** = $\mathbf{V}_{out}/\mathbf{E}$ for the *RL* high-pass filter of Figure 24-3.

5. Determine the cutoff frequencies (in radians per second and in hertz) of the circuit in Figure 24-3. Record the values in Table 24-3.

TABLE 24-3

ω_C	rad/s
f_C	Hz

6. Sketch the straight-line approximations of the frequency responses (A_v in dB versus frequency and θ versus frequency) for the *RL* high-pass filter on Graph 24-2 (page 191).

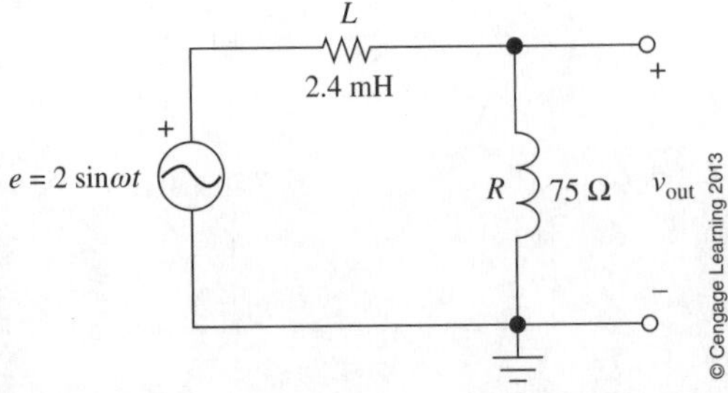

FIGURE 24-3 *RL* high-pass filter.

MEASUREMENTS

The *RC* High-Pass Filter

7. Assemble the circuit shown in Figure 24-2. Connect Ch1 of the oscilloscope to the output of the signal generator and adjust the output to have an amplitude of 2.0 V_p (4.0 $V_{p\text{-}p}$) at a frequency of f = 100 Hz.
8. Use Ch1 as the trigger source and connect Ch2 of the oscilloscope across the resistor R. Measure the amplitude and phase angle (with respect to e) of the sinusoidal output voltage v_{out}. Enter the results in Table 24-4.
9. Increase the frequency of the signal generator to the frequencies indicated in Table 24-4. If necessary, adjust the amplitude of the signal generator to ensure that the output is maintained at 2.0 V_p (4.0 $V_{p\text{-}p}$). Measure the amplitude and phase shift of v_{out} (with respect to e) for each frequency.

TABLE 24-4

f	v_{out}	
	Amplitude	Phase shift
100 Hz		
200 Hz		
400 Hz		
800 Hz		
1 kHz		
2 kHz		
4 kHz		
8 kHz		
10 kHz		

10. Determine the cutoff frequency by adjusting the frequency of the generator until the output voltage has an amplitude of $V_{out} = (0.707)(2.0\ V_p) = 1.41\ V_p$. Record the measured cutoff frequency below.

f_C	

The *RL* High-Pass Filter

11. Construct the circuit shown in Figure 24-3. Connect Ch1 of the oscilloscope to the output of the signal generator and adjust the output to have an amplitude of 2.0 V_p (4.0 $V_{p\text{-}p}$) at a frequency of f = 100 Hz.
12. Use Ch1 as the trigger source and connect Ch2 of the oscilloscope across the inductor L. Measure the amplitude and phase angle (with respect to e) of the sinusoidal output voltage v_{out}. Enter the results in Table 24-5.
13. Increase the frequency of the signal generator to the frequencies indicated in Table 24-5. If necessary, adjust the amplitude of the signal generator to ensure that the output is maintained at 2.0 V_p (4.0 $V_{p\text{-}p}$). Measure the amplitude and phase shift of v_{out} (with respect to e) for each frequency.

TABLE 24-5

f	v_{out}	
	Amplitude	Phase shift
100 Hz		
200 Hz		
400 Hz		
800 Hz		
1 kHz		
2 kHz		
4 kHz		
8 kHz		
10 kHz		

14. Determine the cutoff frequency by adjusting the frequency of the generator until the output voltage has an amplitude of V_{out} = (0.707)(2.0 V_p) = 1.41 V_p. Record the measured cutoff frequency below.

f_C	

CONCLUSIONS

The *RC* High-Pass Filter

15. Use the measurements recorded in Table 24-4 to calculate the magnitude of the gain as a ratio of the amplitudes V_{out}/V_{in}. Calculate the voltage gain in decibels as

$$[A_v]_{dB} = 20 \log \frac{V_{out}}{V_{in}} \tag{24-6}$$

Record the calculated voltage gain and measured phase shift for each frequency in Table 24-6.

TABLE 24-6

f	$A_v = V_{out}/V_{in}$	$[A_v]_{dB}$	θ
100 Hz			
200 Hz			
400 Hz			
800 Hz			
1 kHz			
2 kHz			
4 kHz			
8 kHz			
10 kHz			

16. Plot the data of Table 24-6 on the semilogarithmic scales of Graph 24-1 (next page). Connect the points with the best smooth, continuous curve. (Alternatively, your instructor may have you use Excel to graph your data.) You should find that the actual response is closely approximated by the Bode plot.
17. Compare the measured cutoff frequency f_C of Step 10 to the theoretical value predicted by the transfer function in Step 2.

____

____

The *RL* High-Pass Filter

18. Use the measurements recorded in Table 24-5 to calculate the magnitude of the gain as a ratio of the amplitudes V_{out}/V_{in}. Determine the voltage gain in decibels. Record the calculated voltage gain and measured phase shift for each frequency in Table 24-7.

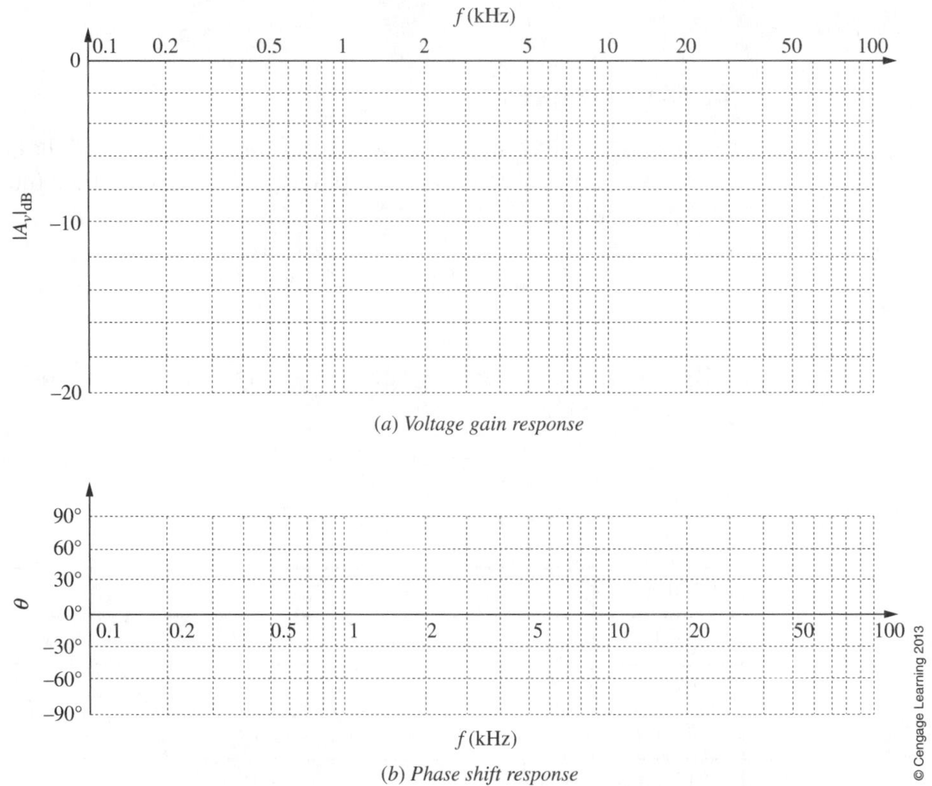

GRAPH 24-1 Frequency response of an *RC* high-pass filter.

TABLE 24-7

f	$A_v = V_{out}/V_{in}$	$[A_v]_{dB}$	θ
100 Hz			
200 Hz			
400 Hz			
800 Hz			
1 kHz			
2 kHz			
4 kHz			
8 kHz			
10 kHz			

19. Plot the data of Table 24-7 on the semilogarithmic scales of Graph 24-2. Connect the points with the best smooth, continuous curve. (Alternatively, your instructor may have you use Excel to graph your data.) You should find that the actual response is closely approximated by the Bode plot.
20. Compare the measured cutoff frequency f_C of Step 14 to the theoretical value predicted by the transfer function in Step 5.

____

____

FOR FURTHER INVESTIGATION AND DISCUSSION

21. Use Multisim or PSpice to simulate the circuit of Figure 24-2. Obtain the frequency response curves for this circuit and compare the results to the actual measurements.
22. Repeat Step 21 for the circuit of Figure 24-3.

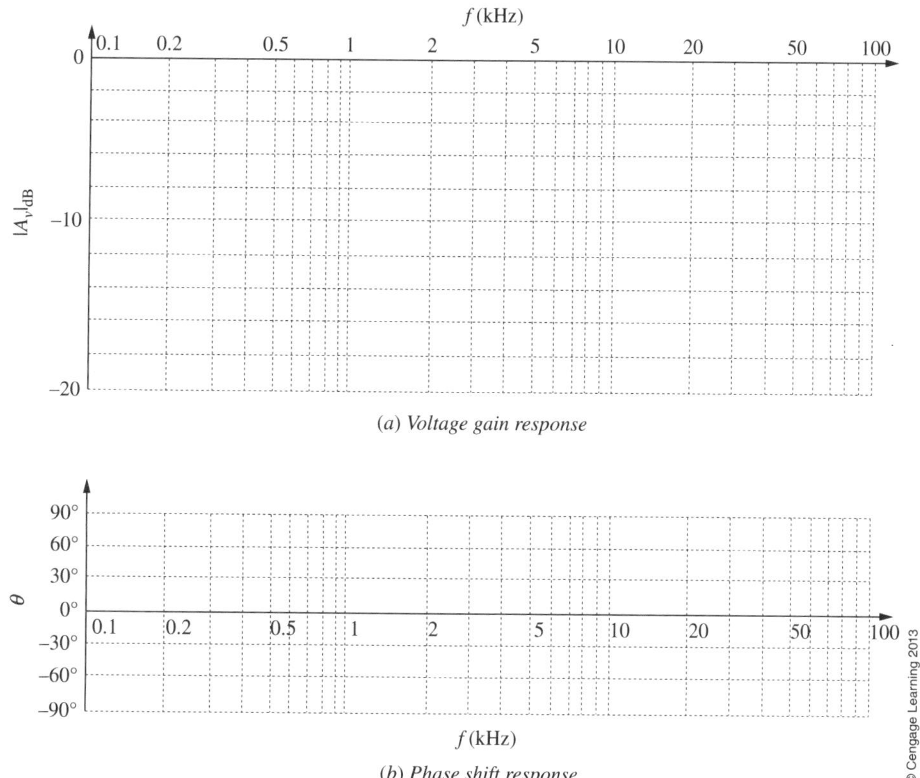

GRAPH 24-2 Frequency response of an *RL* high-pass filter.

Name ______________________

Date ______________________

Class ______________________

LAB 25

Band-Pass Filter

OBJECTIVES

After completing this lab, you will be able to

- calculate the cutoff frequencies of a band-pass filter by examining the individual low-pass and high-pass stages of a filter,
- derive the transfer function for each stage of a band-pass filter,
- sketch the Bode plot of a band-pass filter from the transfer function of the individual stages,
- explain why the slope of the voltage gain response is 20 dB/decade on each side of the cutoff frequencies.

EQUIPMENT REQUIRED

- ☐ Dual-trace oscilloscope
- ☐ Signal generator (sinusoidal function generator)
- ☐ Digital multimeter (DMM)

Note: Record this equipment in Table 25-1.

COMPONENTS

- ☐ Resistors: 330-Ω (2) (1/4-W carbon, 5% tolerance)
- ☐ Capacitors: 0.047-μF, 0.47-μF (10% tolerance)

EQUIPMENT USED

TABLE 25-1

Instrument	Manufacturer/Model No.	Serial No.
Oscilloscope		
Signal generator		
DMM		

TEXT REFERENCE

Section 22.6 BAND-PASS FILTER

DISCUSSION

Band-pass filters permit a range of frequencies to pass from one stage to another. Many different types of band-pass filters are used throughout electronics. For instance, the typical television receiver uses several stages of filtering to achieve a bandwidth of 6 MHz, while an AM receiver has circuitry which restricts the bandwidth to only 10 kHz. Although band-pass filters can be very elaborate, depending on the application, they can also be constructed simply by combining a low-pass filter and a high-pass filter as shown in Figure 25-1. Since expense and other design difficulties make *RL* filters impractical, *RC* filters are used almost exclusively.

The band-pass filter of Figure 25-1 has two cutoff frequencies as determined by the cutoff frequencies of the individual stages. The frequency below which the high-pass filter attenuates the signal is called the *lower cutoff frequency* f_1. The frequency at which the low-pass filter begins to attenuate is called the *upper cutoff frequency* f_2. As expected, the difference between the two frequencies is the *bandwidth BW*. In order for the circuit to operate predictably, the cutoff frequencies should be separated by at least one *decade* and the first stage must be the high-pass circuit.

The lower cutoff frequency ω_1 in radians per second is found as

$$\omega_1 = \frac{1}{\tau_1} = \frac{1}{R_1 C_1} \tag{25-1}$$

with a corresponding frequency f_1 in hertz,

$$f_1 = \frac{\omega_1}{2\pi} = \frac{1}{2\pi R_1 C_1} \tag{25-2}$$

The upper cutoff frequency ω_2 in radians per second is

$$\omega_2 = \frac{1}{\tau_2} = \frac{1}{R_2 C_2} \tag{25-3}$$

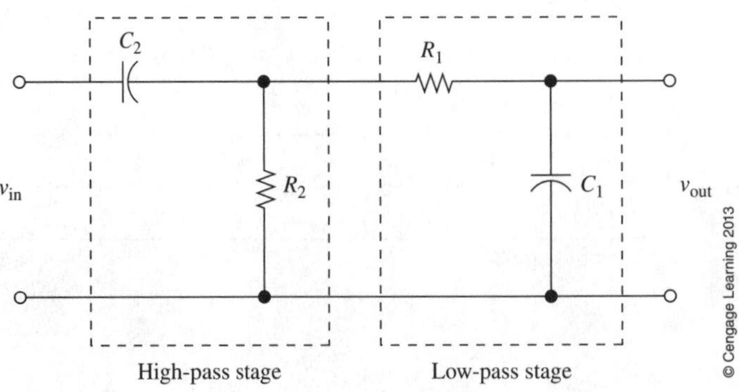

FIGURE 25-1 Simple band-pass filter.

with a corresponding frequency f_2 in hertz,

$$f_2 = \frac{\omega_2}{2\pi} = \frac{1}{2\pi R_2 C_2} \qquad (25\text{-}4)$$

The bandwidth of the filter is determined as

$$BW = f_2 - f_1 \qquad (25\text{-}5)$$

CALCULATIONS

1. Calculate and record the cutoff frequency (in radians per second and in hertz) for the high-pass stage in the circuit of Figure 25-2.

ω_1	rad/s
f_1	Hz

2. Calculate and record the cutoff frequency (in radians per second and in hertz) for the low-pass stage in the circuit of Figure 25-2.

ω_2	rad/s
f_2	Hz

3. Calculate and record the bandwidth of the filter circuit of Figure 25-2.

BW	Hz

4. From the calculations of Steps 1 and 2 sketch the Bode plots for the band-pass filter of Figure 25-2. Use the semilogarithmic scales of Graph 25-1 (page 198).

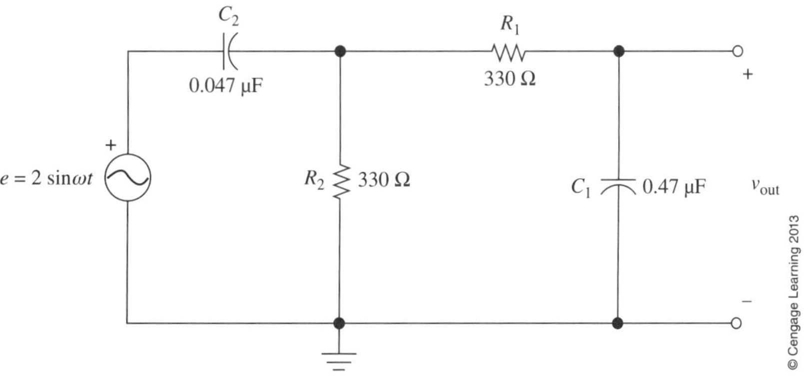

FIGURE 25-2 Band-pass filter.

MEASUREMENTS

5. Assemble the circuit shown in Figure 25-2. Connect Ch1 of the oscilloscope to the output of the signal generator and adjust the output to have an amplitude of 2.0 V_p (4.0 $V_{p\text{-}p}$) at a frequency of f = 100 Hz.
6. Use Ch1 as the trigger source and connect Ch2 of the oscilloscope across the output of the circuit. Measure the amplitude and phase angle of the output voltage v_{out} (with respect to e). Enter the results in Table 25-2.
7. Increase the frequency of the signal generator to the frequencies indicated in Table 25-2. Due to loading effects of the circuit, it will be necessary to readjust the amplitude of the signal generator at each frequency to ensure that the output is maintained at 2.0 V_p (4.0 $V_{p\text{-}p}$). Measure and record the amplitude and phase shift of the output voltage v_{out} (with respect to e) for each frequency.

TABLE 25-2

f	v_{out}	
	Amplitude	Phase shift
100 Hz		
200 Hz		
400 Hz		
800 Hz		
1 kHz		
2 kHz		
4 kHz		
8 kHz		
10 kHz		
20 kHz		
40 kHz		
80 kHz		
100 kHz		

8. For which frequency f in Table 25-2 is the output voltage a maximum? Adjust the generator to provide an output of 2.0 V_p (4.0 $V_{p\text{-}p}$) at this frequency. While observing the oscilloscope, adjust the the generator frequency until voltage V_{out} is at the maximum value. Measure and record the center frequency f_0 and the corresponding amplitude V_{out}.

f_0	

9. Decrease the frequency until the output voltage V_{out} is reduced to 0.707 of the maximum value found in Step 8. Record the lower half-power frequency f_1 and the corresponding output voltage V_{out}. (Ensure that the output of the signal generator is at 2.0 V_p.)

f_1	

10. Increase the frequency above the center frequency until the output voltage V_{out} is again reduced to 0.707 of the maximum value found in Step 8. Record the upper half-power frequency f_2 and the corresponding resistor voltage V_{out}. (Ensure that the output of the signal generator is at 2.0 V_p.)

f_2	

CONCLUSIONS

11. For each voltage measurement in Table 25-2, determine the voltage gain as both a ratio of amplitudes $A_v = v_{out}/E$ and in decibels as

$$[A_v]_{dB} = 20 \log \frac{V_{out}}{E} \tag{25-6}$$

Record the calculated voltage gain and measured phase shift for each frequency in Table 25-3.

12. Plot the data of Table 25-3 on the semilogarithmic scale of Graph 25-1. Connect the points with the best smooth, continuous curve. (Alternatively, your instructor may have you use Excel to graph your data.)
13. Compare the measured cutoff frequencies of Steps 9 and 10 to the theoretical frequencies determined in Steps 1 and 2.

____

____

14. How do the actual voltage gain and phase shift responses of the output voltage compare to the predicted responses?

____

____

____

TABLE 25-3

f	$A_v = V_{out}/E$	$[A_v]_{dB}$	θ
100 Hz			
200 Hz			
400 Hz			
800 Hz			
1 kHz			
2 kHz			
4 kHz			
8 kHz			
10 kHz			
20 kHz			
40 kHz			
80 kHz			
100 kHz			

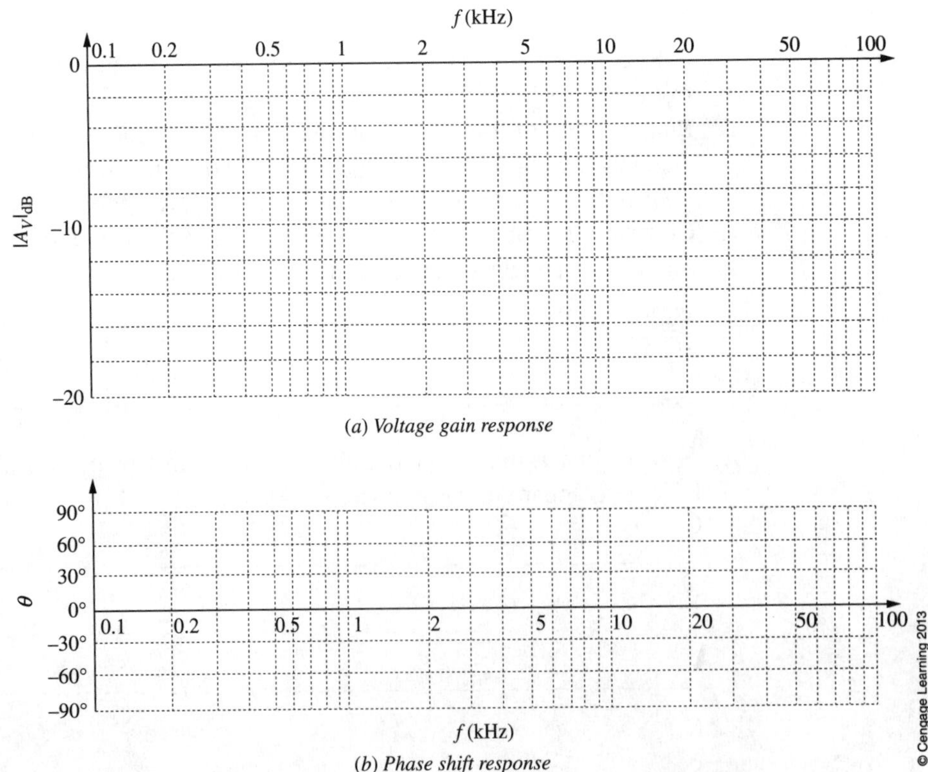

GRAPH 25-1 Frequency response of a simple band-pass filter.

Name ______________________

Date ______________________

Class ______________________

LAB 26

Voltage, Current, and Power in Balanced Three-Phase Systems

OBJECTIVES

After completing this lab, you will be able to

- verify line and phase voltage relationships for a Y-load,
- verify line and phase current relationships for a Δ-load,
- verify the single-phase equivalent method of analysis,
- measure power in a three-phase system.

EQUIPMENT REQUIRED

- ☐ ac ammeters (DMMs with 2-A ranges are adequate)
- ☐ Digital multimeters (DMMs)
- ☐ Single-phase wattmeter (two required)

POWER SUPPLY

- ☐ Three-phase 120/208 V (60 Hz), preferably variable such as by means of a three-phase autotransformer

COMPONENTS

- ☐ Resistors: 100-Ω, 200-W (3), 250-Ω, 200-W (3)
- ☐ Capacitors: 10 μF, nonelectrolytic, rated for operation at 120 Vac (three required)

PRE-STUDY (OPTIONAL)

For Multisim based pre-study simulations, go to the Student Premium Website and choose *Three Phase Pre-Study*. You might also want to use *CircuitSim 24-1* and *24-2*.

EQUIPMENT USED

TABLE 26-1

Instrument	Manufacturer/Model No.	Serial No.
Single-phase wattmeters		
ac Ammeter or DMM		
DMMs		

TEXT REFERENCE

Section 24.1 THREE-PHASE VOLTAGE GENERATION
Section 24.3 BASIC THREE-PHASE RELATIONSHIPS
Section 24.5 POWER IN A BALANCED SYSTEM
Section 24.6 MEASURING POWER IN THREE-PHASE CIRCUITS

DISCUSSION

Y-Loads. For a Y-load, Figure 26-1, line-to-line voltages are $\sqrt{3}$ times line-to-neutral voltages, and each line-to-line voltage leads its corresponding line-to-neutral voltage by 30°. Thus,

$$\mathbf{V}_{ab} = \sqrt{3}\ \mathbf{V}_{an}\angle 30° \qquad (26\text{-}1)$$

Current in the neutral of a balanced system (if a neutral line is present), is zero.

Δ-Loads. The magnitude of line current for a balanced Δ-load, Figure 26-2, is $\sqrt{3}$ times the magnitude of the phase current and each line current, lags its corresponding phase current by 30°. Thus,

$$\mathbf{I}_{a} = \sqrt{3}\ \mathbf{I}_{ab}\angle -30° \qquad (26\text{-}2)$$

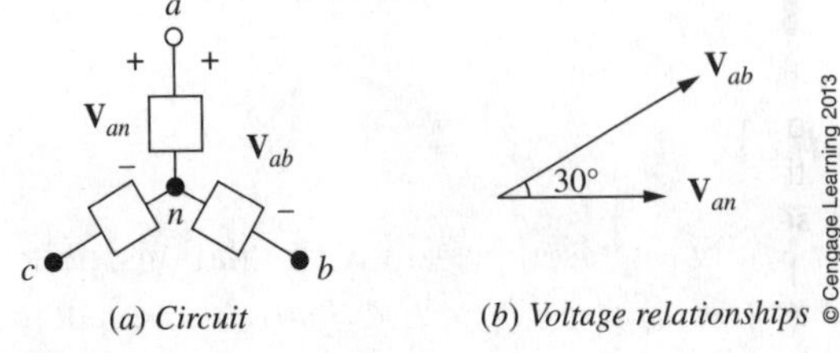

FIGURE 26-1 A balanced Y-load.

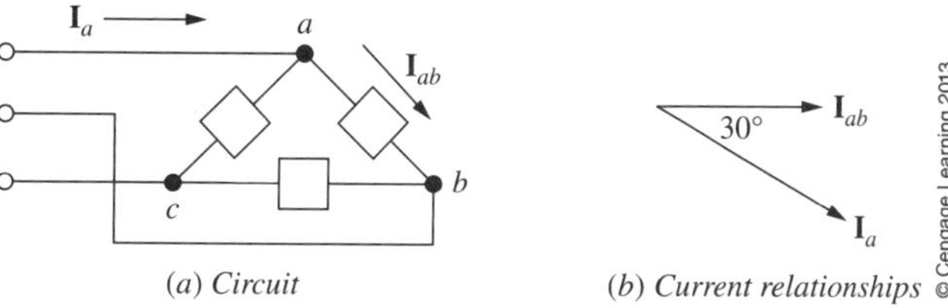

FIGURE 26-2 A balanced Δ-load.

Equivalent Y- and Δ-Loads. For purposes of analysis, a balanced Δ-load may be replaced by an equivalent balanced Y-load with

$$\mathbf{Z}_{\mathrm{Y}} = \mathbf{Z}_{\Delta}/3 \tag{26-3}$$

Single-Phase Equivalent. Since all three phases of a balanced system are identical, any one phase can be taken to represent the behavior of all. This permits you to reduce a circuit to its *single-phase equivalent.*

Power in Three-Phase Systems. Power in a balanced three-phase system is three times the power to one phase. Power to one phase is given by

$$P_{\phi} = V_{\phi} I_{\phi} \cos\theta_{\phi} \tag{26-4}$$

where V_{ϕ} and I_{ϕ} are the magnitudes of the phase voltage and current and θ_{ϕ}, the angle between them, is the angle of the phase load impedance. This formula applies to both Y- and Δ-loads. Total power is three times this.

An alternate power formula based on line voltages and currents is

$$P_{\mathrm{T}} = \sqrt{3}\, V_{\mathrm{L}} I_{\mathrm{L}} \cos\theta_{\phi} \tag{26-5}$$

where V_{L} and I_{L} are the magnitudes of the line-to-line voltages and line currents, respectively. This formula applies to both Y- and Δ-loads. Note that the angle in this formula is the angle of the load impedance—it is not the angle between $\mathbf{V}_{\mathrm{L}}$ and $\mathbf{I}_{\mathrm{L}}$.

Measuring Power in Three-Phase Systems. For a three-wire load, you need only two wattmeters, while for a four-wire load, you need three wattmeters, Figure 26-3. Note however, the use of individual wattmeters as illustrated here is giving way to the use of integrated, multifunction meters (recall Figure 17-23 of the textbook) that incorporate power/energy measurement for all three phases in a single package.

Safety Note

In this lab, you will be working with 120/208 Vac. Such voltages are dangerous, and you must be aware of and observe safety precautions. Familiarize yourself with your laboratory's safety practices. **Always ensure that power is off when you are assembling, changing, or otherwise working on your circuit.** If possible, use a variable ac power supply (such as a three-phase variable autotransformer) as the source. Start at zero volts and gradually increase the voltage while watching the meters for signs of trouble. In any event, have your instructor check your circuit before you energize it. **USE EXTREME CAUTION IN THIS LAB.**

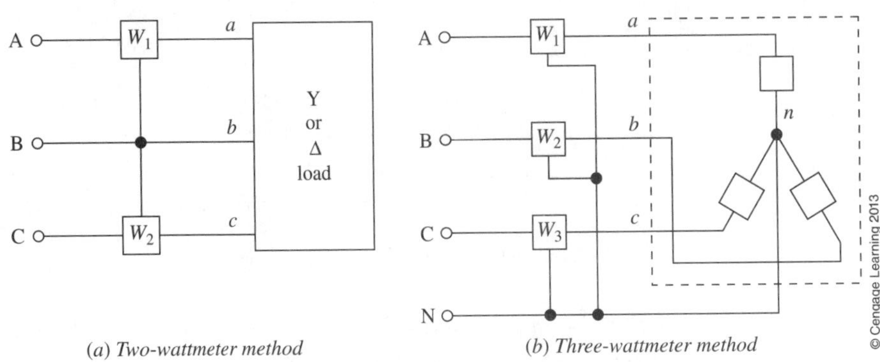

FIGURE 26-3 Measuring power in three-phase systems.

MEASUREMENTS

PART A: Voltages and Currents for a Y-Load

1. With power off, set up the circuit of Figure 26-4. Use dedicated ac ammeters or DMMs with a current range of 2 A or more to measure current. Observe all safety precautions—see General Safety Notes.
   a. Set E_{AN} to 120 V (or as close as you can get it) using a DMM, then measure all line-to-neutral and line-to-line voltages at the load. Measure all currents (line and neutral). Record in Table 26-2. Do your measurements confirm the magnitude relationship of Equation 26-1?

TABLE 26-2

V_{an}	V_{bn}	V_{cn}	V_{ab}	V_{bc}	V_{ca}	I_a	I_b	I_c	I_n

   b. Turn power off and remove the neutral conductor from between n and N. Measure the line currents again. Are they the same as in Test 1(a)?
   c. Using the measured source voltage and load resistance, compute the magnitude of the line current and compare to the measured value.

General Safety Notes

1. With power off, assemble your circuit and have your lab partner double check it.
2. Have your instructor check the circuit before you energize it. If you are using a Powerstat or other variable ac source, gradually increase voltage from zero, watching the meters for signs of trouble.
3. Turn power off before changing your circuit for the next test.

Check with your instructor for specific safety instructions.

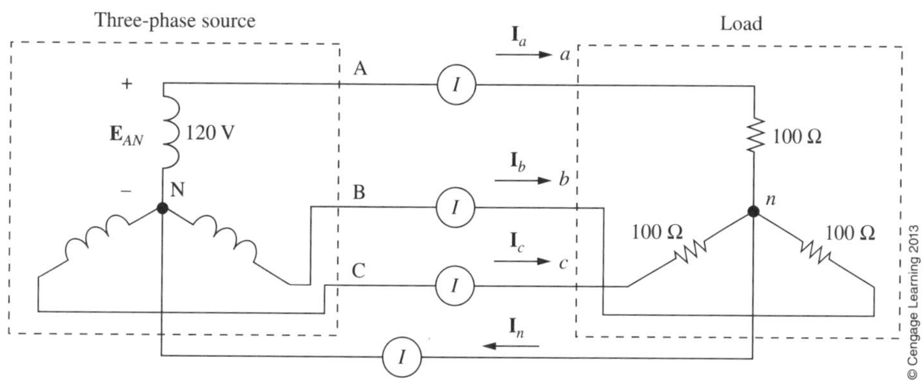

FIGURE 26-4 Circuit for Test 1. Each resistor is rated 200 W.

PART B: Power to a Y-Load

2. With power off, add wattmeters as in Figure 26-5. Adjust E_{AN} to the same value you used in Part A. (Check currents and note if any have changed. They should not have.) Record wattmeter readings.

   W_1 = ___________ W_2 = ___________

3. Using circuit analysis, determine P_T. Now sum the wattmeter readings. How well do the results agree?
4. Using circuit analysis, determine what wattmeter W_1 should read. Compare it to the actual reading. Repeat for wattmeter W_2.

PART C: Power to a Y-Load (Continued)

5. De-energize the circuit and add capacitors as in Figure 26-6. Set the voltage as in Part A. Measure line current and power and record.

   I = ___________ W_1 = ___________ W_2 = ___________

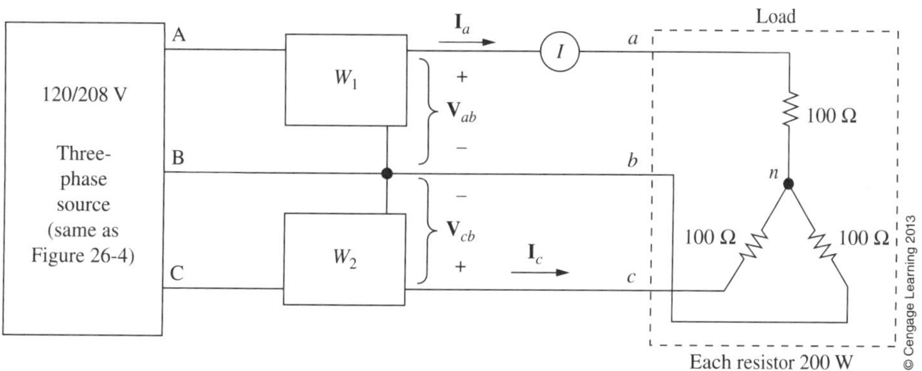

FIGURE 26-5 Circuit for Test 2 (for details of the three-phase source, see Figure 26-4).

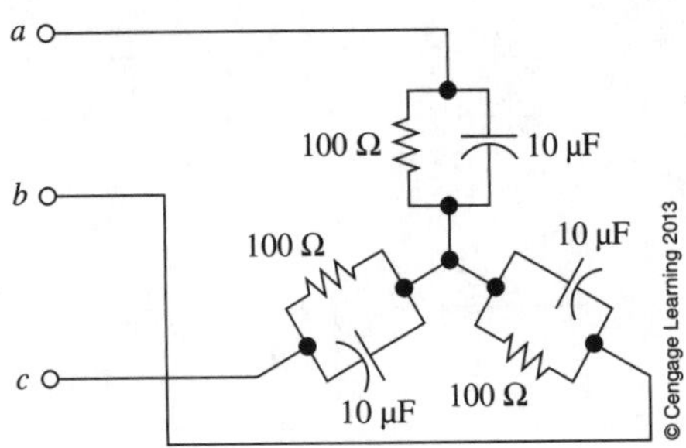

FIGURE 26-6 Load for Test 5.

a. Although the wattmeter readings change, total power should remain unchanged. Why? Verify by using circuit analysis techniques to compute P_T.
b. Using circuit analysis techniques, compute the reading of W_1. Compare to the measured value. Repeat for W_2.

PART D: Currents for a Δ-Load

6. With power off, assemble the circuit of Figure 26-7, using 250-Ω resistors. Measure line current I_a and phase current I_{ab}. Do your measurements confirm the magnitude relationship of Equation 26-2?

I_a = ____________ I_{ab} = ____________

Part E: Power to a Δ-Load

7. De-energize the circuit and add wattmeters as in Figure 26-3(a). Set the voltage as in Part A. Measure power and record.

W_1 = ____________ W_2 = ____________

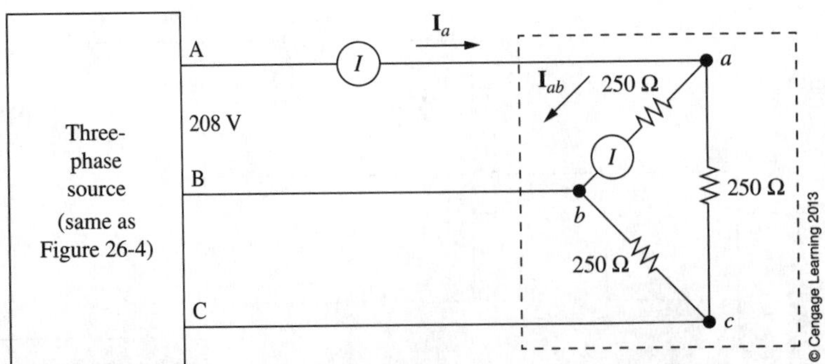

FIGURE 26-7 Circuit for Test 6 (for three-phase source, see Figure 26-4).

8. Using circuit analysis techniques, analyze the circuit of Figure 26-7, computing line current, total power, and the reading of each wattmeter. Compare your computed results to the measured results. How well do they agree?

PART F: The Single-Phase Equivalent

9. Consider the three-phase loads of Figure 26-8(a). (Do not assemble this circuit.) Each resistor of the Y is 250 Ω and each resistor of the Δ is 300 Ω. Source voltage is 120-V, line-to-neutral.
   a. Convert the Δ-load to a Y-equivalent, then convert the circuit to its single-phase equivalent. Sketch as Figure 26-8(b).
   b. Assemble the single-phase equivalent. Set the source voltage as in Part A and measure line current.

   E_{AN} = ____________ I_A = ____________

   c. Analyze the single-phase equivalent and compare calculated current I_A to that measured in (b).

PROBLEMS

10. Assume each resistor of the Δ-load of Figure 26-8(a) is replaced with a 10-F capacitor.
    a. Convert the circuit to its single-phase equivalent and sketch.
    b. Calculate the magnitude of the line current I_A.
    c. Calculate the magnitude of the phase current in the Δ-load.
    d. Calculate total power to the combined load.

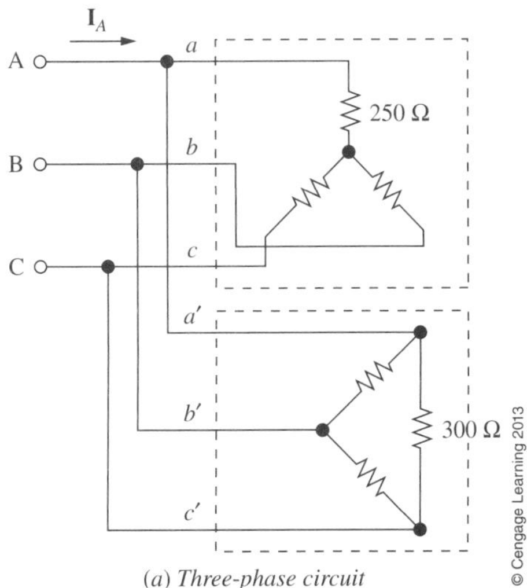

(a) Three-phase circuit (b) Single-phase equivalent

FIGURE 26-8 Circuit for Test 9.

COMPUTER ANALYSIS

11. Using Multisim or PSpice, verify the currents computed in Question 10 and the current measured in Part F.

FOR FURTHER INVESTIGATION AND DISCUSSION

12. Using Multisim, click *Place, Component, Sources, THREE_PHASE_WYE* and place a three-phase source on your screen. Right-click your menu bar and ensure that *Instruments* is selected. Select and place both a two-channel scope and a four-channel scope on your screen. (Use generic scopes, not Tektronix or Agilent.)
    a. Using the floating inputs of the two-channel scope, connect to display voltage v_{an} on Ch A and voltage v_{ab} on Ch B. Print out waveforms. (If you do not have a printer, carefully sketch.)
    b. Using cursors, measure the displacement between v_{ab} and v_{an}.
    c. Using a Multisim DMM, measure V_{an} and V_{ab}.
    d. Using the four-channel scope, connect to display voltages v_{ab}, v_{bc}, and v_{ca}. Print out waveforms. (If you do not have a printer, carefully sketch.)

Write a short report describing how your studies here confirm the theory discussed in the textbook.

13. Using PSpice, set up the three-phase source of Figure 26-4 and investigate three-phase voltage waveforms. (Since you need time varying waveforms here, use VSIN as your source if you are using PSpice and select Transient Analysis.) Use the following to guide your analysis:
    a. Plot waveforms for phase voltages v_{AN}, v_{BN}, and v_{CN}.
    b. Repeat for line voltages v_{AB}, v_{BC}, and v_{CA}.
    c. Plot v_{AN} and v_{AB} on the same graph; then, using the cursor, measure values and verify the magnitude and phase relationships of Equation 26-1.

Write a short report confirming how your study verifies the results discussed in the text.

Name ______________________

Date ______________________

Class ______________________

LAB 27 The Iron-Core Transformer

Safety Note

In parts of this lab, you will be working with 120 Vac. This is a dangerous voltage and you must be aware of and observe safety precautions. Familiarize yourself with your lab's safety practices and use extreme caution at all times. Always ensure that power is off when you are assembling, changing, or otherwise working on circuits with dangerous voltages.

OBJECTIVE

After completing this lab, you will be able to
- verify the turns ratio and phase relationships for a transformer,
- verify the concept of reflected impedance,
- determine the frequency response of an audio transformer,
- measure the regulation of a power transformer.

EQUIPMENT REQUIRED

☐ Oscilloscope
☐ Digital multimeter (DMM)
☐ Signal or function generator

COMPONENTS

☐ Resistors: 10-Ω, 100-Ω (each 1/4 W), 10-Ω, 25-W
☐ Audio transformer (Hammond 145F or equivalent)
☐ 120/12.6-V filament transformer (Hammond 167L12 or equivalent)

PRE-STUDY (OPTIONAL)

Interactive simulations *CircuitSim 23-1, 23-2, 23-5,* and *23-7* may be used as pre-study. From the Student Premium Website, select *Learning Circuit Theory Interactively on Your Computer*, download and unzip *Ch 23* (if you have not already done so), and run the corresponding *Sim* files.

EQUIPMENT USED

TABLE 27-1

Instrument	Manufacturer/Model No.	Serial No.
Oscilloscope		
DMM		
Signal or function generator		

TEXT REFERENCE

Section 23.2 IRON-CORE TRANSFORMERS: THE IDEAL MODEL
Section 23.3 REFLECTED IMPEDANCE
Section 23.8 VOLTAGE AND FREQUENCY EFFECTS

DISCUSSION

We look first at the ideal transformer. An ideal transformer, Figure 27-1(a), is one that is characterized by its turns ratio

$$a = N_p/N_s \qquad (27\text{-}1)$$

where N_p and N_s are its primary and secondary turns, respectively. Voltage and current ratios are given by

$$\mathbf{V}_p/\mathbf{V}_s = V_p/V_s = a \qquad (27\text{-}2)$$

$$\mathbf{I}_p/\mathbf{I}_s = I_p/I_s = 1/a \qquad (27\text{-}3)$$

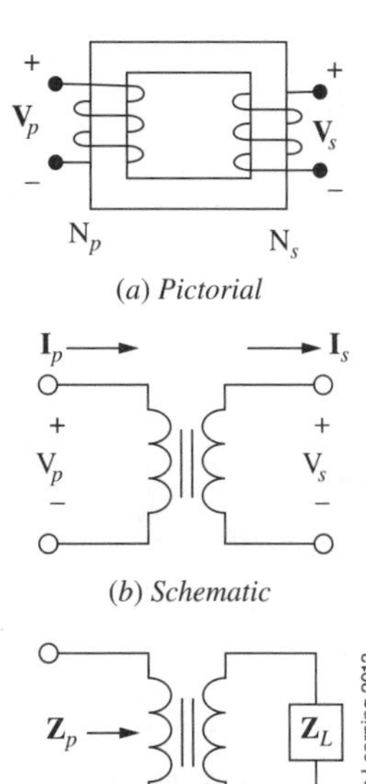

FIGURE 27-1 The ideal transformer.

Reflected Impedance. Load impedance $\mathbf{Z}_L$, Figure 27-1(c), is reflected into the primary as

$$\mathbf{Z}_p = a^2 \mathbf{Z}_L \qquad (27\text{-}4)$$

All impedances in the secondary (whether part of the load or not) are reflected in this fashion.

Phasing. Depending on the relative direction of windings, the secondary voltage is either in-phase or 180° out-of-phase with respect to the primary—see Practical Note 1.

Real Transformers. Real transformers differ from the ideal in that they have primary and secondary winding resistances, leakage flux, core losses, and a few other nonideal characteristics as described in Sections 23.6 to 23.8 of the text. In this lab, you will observe some of these—you will look in particular at the regulation of a power transformer and the bandwidth limitations of an audio transformer.

Voltage Ratios. The voltage ratio of an iron-core transformer under no load is the same as its turns ratio. However (because of internal voltage drops), the ratio under load is different—see Practical Note 2.

MEASUREMENTS

PART A: Turns Ratio

1. Consider Figure 27-2. Observing suitable safety precautions, apply full rated voltage to the transformer.

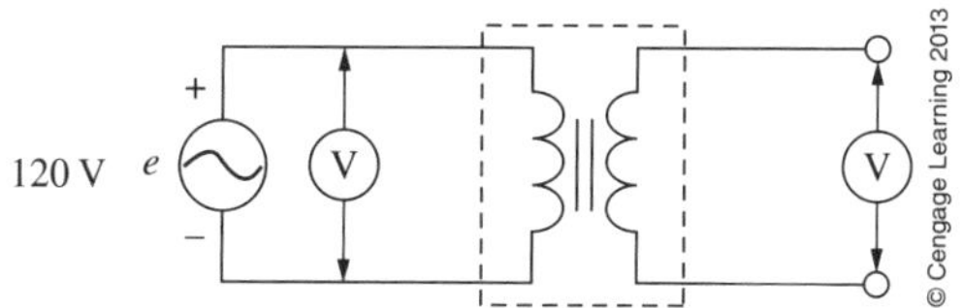

FIGURE 27-2 Circuit for Test 1. Use a filament transformer (or equivalent) with a 120-V primary and a low-voltage (about 20 V or less) secondary.

Practical Notes

1. Small transformers usually have color-coded leads to identify corresponding (dotted) terminals, while power transformers often use letter designations H_1, X_1, etc.—check the data sheet or nameplate of your transformer for details.
2. Some data sheets list the ratio of primary voltage to secondary voltage under load, some under no load, and some list both. For example, the Hammond 167L12 suggested for this lab has a no-load specification of 115V/13.8V and a full-load specification of 115V/12.6V.

a. Measure primary and secondary voltages.

V_p = __________; V_s = __________

Determine the turns ratio using Equation 27-2.

$a_{(\text{measured})}$ = __________

b. Determine the turns ratio from its no-load data sheet and compare it to that of Test 1(a).

$a_{(\text{Nameplate})}$ = __________

PART B: Phase Relationships

2. Using the audio transformer, assemble the circuit of Figure 27-3. Set the source to a 1-kHz sine wave with an amplitude of 8 V (i.e., 16 $V_{p\text{-}p}$). Mark the transformer terminals for reference in determining phase relationships.
   a. Measure the magnitude and phase of v_{cd} relative to the primary voltage and sketch in Figure 27-4. Now reverse the secondary scope leads and sketch v_{dc}. Based on these observations, add the missing dot to the transformer of Figure 27-3. Explain how you determined the position of this dot.

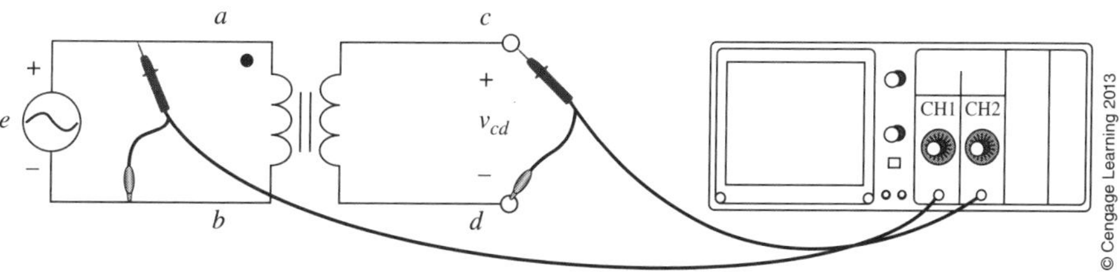

FIGURE 27-3 Circuit for Test 2 (use the audio transformer).

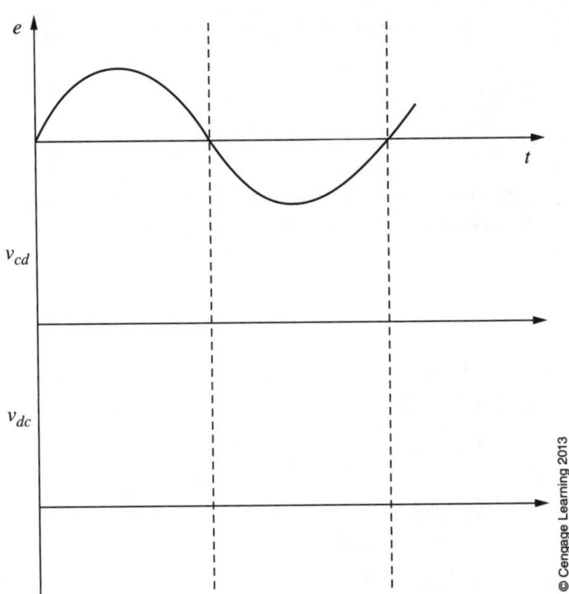

FIGURE 27-4 Waveforms for Test 2(a).

b. Based on the measurements of Test 2(a), what is the turns ratio for this transformer?

$a =$ ________

PART C: Reflected Impedance

3. Measure the 10-Ω sensing resistor and the 100-Ω load resistor, then assemble the circuit of Figure 27-5, using the audio transformer. Connect probes as shown.

$R_{sen} =$ ________ $R_L =$ ________

a. Using the oscilloscope, set the input voltage V_p of the transformer to 16-V peak-to-peak, 1 kHz.

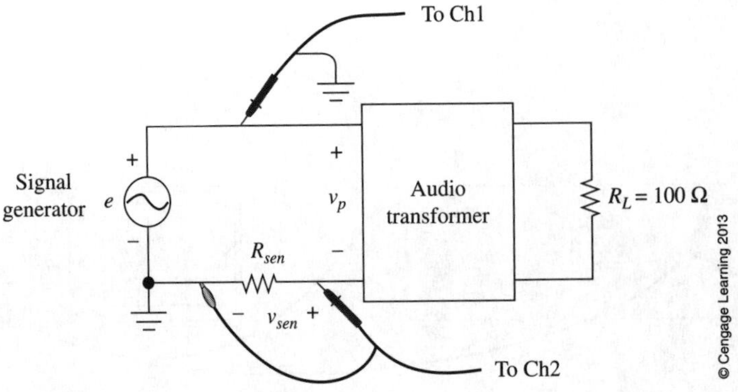

FIGURE 27-5 Circuit for Test 3.

b. Select Ch2 and measure the voltage across the sensing resistor. From this, calculate the primary current.

V_{sen} = __________ I_p = __________

c. Using the results of Test 3(a) and (b), determine the magnitude of the input impedance to the transformer.

$Z_{in(measured)}$ = __________

d. For an ideal transformer, input impedance is $Z_p = a^2Z_L$. However, real transformers have winding resistance and leakage reactance. Neglecting leakage reactance, we get the circuit of Figure 27-6. The winding resistance can now be reflected into the primary along with the load resistance. Adding to this, the primary winding resistance, we get (approximately) an input impedance Z_{in} of

$$Z_{in} = R_p + a^2(R_s + R_L) \qquad (27\text{-}5)$$

Measurements. Measure primary and secondary winding resistances R_p and R_s using an ohmmeter, then use Equation 27-5 to compute the input impedance for the transformer. Compare to the value determined in (c), that is, calculate the percent difference. If they differ, what are the likely sources of the difference?

Z_{in} = __________

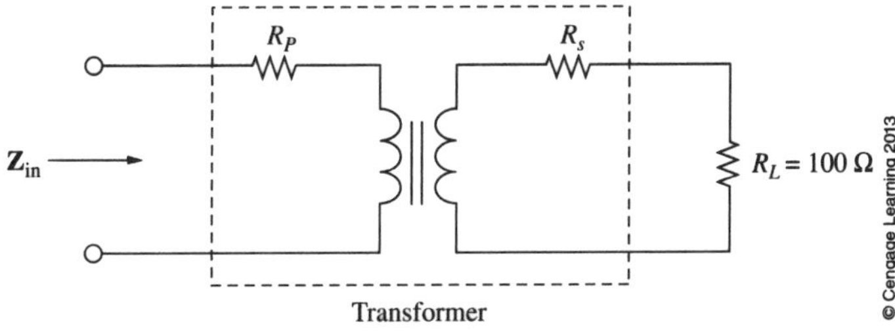

FIGURE 27-6 Transformer circuit with winding resistance included.

PART D: Frequency Response of a Transformer

4. a. Using the audio transformer, reconnect the circuit of Figure 27-3, but add a 100-Ω load resistor between points *c* and *d*. Maintain the input voltage constant at 6 V (i.e., 12 $V_{p\text{-}p}$) and measure the output voltage and record in Table 27-2.
   b. Compute the ratio of output voltage to input voltage over the range of frequencies tested and plot as Figure 27-7.
   c. From the observed data, what conclusion can you draw about the validity of the ideal model?

TABLE 27-2

f	V_{out}	V_{out}/V_{in}
100 Hz		
500 Hz		
1 kHz		
5 kHz		
10 kHz		
15 kHz		
20 kHz		

PART E: Load Voltage Regulation

5. Ideally, a transformer will deliver constant output voltage. In reality, due to internal voltage drops, the output voltage falls as load current increases. In this test, we look at how badly the transformer voltage drops.
   a. Omitting the primary-side voltmeter and observing suitable safety precautions, set up the circuit of Figure 27-2 and apply full rated voltage to the transformer.
   b. With a meter, measure the open-circuit (no-load) voltage on the 12.6-V winding.

   V_{NL} = ________

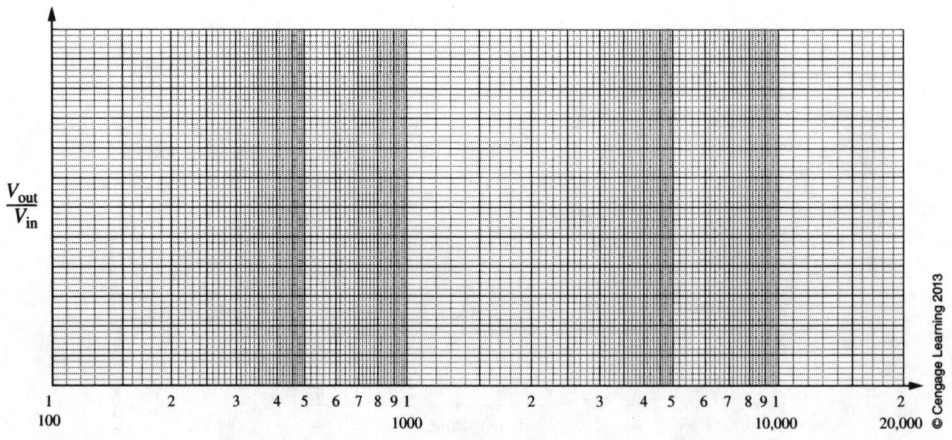

FIGURE 27-7 Frequency response for an audio transformer.

c. Connect a load resistor that will draw full-rated current (or nearly full-rated current). (If you use the Hammond 167L12 as recommended, you will need about 6 Ω, capable of dissipating 25 W. Since this is not a standard resistor, use a 10-Ω resistor instead. You will easily observe the regulation.) Measure output voltage under load.

V_{FL} = __________

d. Compute the regulation of the transformer using the formula

$$\text{Regulation} = \left(\frac{V_{NL} - V_{FL}}{V_{FL}}\right) \times 100\%$$

PROBLEMS

6. An ideal transformer with a turns ratio of $a = 5$ has a load of $\mathbf{Z}_L = 10\ \Omega\ \angle 30°$. Source voltage is $\mathbf{E} = 120\ \text{V}\ \angle 0°$.
   a. Compute the load voltage and load current as in Example 23-4 of the text.
   b. Using Equation 27-3, compute source current.
   c. Compute the input impedance $\mathbf{Z}_p$.
   d. Using $\mathbf{Z}_p$, compute source current and compare to the result obtained in (b).
7. A certain load requires a voltage that does not exceed 124 V or drop below 117 V. Two transformers are available, each with a no-load voltage of 120 V. Transformer 1 has a rated maximum regulation of ±3 percent of its no-load voltage and Transformer 2 is rated 120 V ±3 V. Is either transformer suitable for the application? Perform an analysis to find out.

COMPUTER ANALYSIS

8. Using Multisim or PSpice and an ideal transformer with turns ratio $a = 5$ and $E = 120$ V, set up and emulate the circuit of Figure 27-3. (Omit the oscilloscope.) (PSpice users: Use source VSIN. Also place a large value resistor [say 100kΩ] across the secondary terminals of the transformer since PSpice does not like floating nodes.) Investigate waveforms, phase relationships, and transformation ratios. Specifically, use the cursor to measure voltage magnitudes, and in your analysis, use these to verify the turns ratio. Be sure to get printouts for both the in-phase and out-of-phase cases.

Ideal Transformers in Computer Analysis

Multisim: Use the virtual transformer (found in the BASIC_VIRTUAL bin of the *Basic* parts bin). Double click its symbol and set parameters as noted in Section 23.12 of the text to make it correspond to the ideal model. Note that Multisim requires the use of a ground on both sides of the transformer.

PSpice: Use XFRM_LINEAR. (See boxed note in Section 23.12 of the text to refresh your memory on how to set it up to represent an ideal iron-core transformer.)

9. Consider a 10:1 iron-core transformer with $R_p = 16\ \Omega$, $L_p = 100$ mH, $R_s = 0.16\ \Omega$, $L_s = 1$ mH, and a load consisting of a 10-μF capacitor in parallel with a 4-Ω resistor. The source is 120 V∠0°.
   a. Using Multisim or PSpice, solve for the load voltage. (This is full-load voltage.) Now solve for no-load voltage, then compute regulation.
   b. Solve the preceding problem manually (with your calculator) and compare results.

FOR FURTHER INVESTIGATION AND DISCUSSION

Using the equivalent circuit of Figure 23-41 of the text and Multisim or PSpice, investigate the frequency response of an audio transformer. Use the following parameter values: $R_p = 1\ \Omega$, $L_p = 0.125$ mH, $R_s = 0.04\ \Omega$, $L_s =$ 0.005 mH, $R_C = 500\ \Omega$, $L_m = 0.1$ H, turns ratio $a = 5$, and a speaker load of 8 Ω, purely resistive. (Omit the stray capacitances). Use a 1-Vac source and run a frequency response curve from 10 Hz to 30 kHz. Submit a printout of your circuit plus the response curve. Write a short analysis of your investigation, describing in your own words the significance of what the curve shows in terms of the transformer's impact on sound quality in an audio system. Other points to cover include 1) over what range does the ideal model apply and why, and 2) what causes the curve to fall off at each end of the frequency spectrum?

Name ______________________

Date ______________________

Class ______________________

LAB 28

Mutual Inductance and Loosely Coupled Circuits

OBJECTIVE

After completing this lab, you will be able to

- determine mutual inductance experimentally,
- measure mutual voltage and verify theoretically,
- compare calculated and measured results for coupled circuits.

EQUIPMENT REQUIRED

- ☐ Oscilloscope, dual channel
- ☐ Signal or function generator
- ☐ Inductance meter or bridge

COMPONENTS

- ☐ Loosely coupled transformer
- ☐ Resistor: 10-Ω, 1/4-W
- ☐ Capacitors: 1-μF, nonelectrolytic (two required)

PRE-STUDY (OPTIONAL)

Interactive simulations *CircuitSim 23-13* and *23-14* may be used as pre-study. From the Student Premium Website, select *Learning Circuit Theory Interactively on Your Computer*, download and unzip *Ch 23* (if you have not already done so), and run the corresponding *Sim* files.

EQUIPMENT USED

TABLE 28-1

Instrument	Manufacturer/Model No.	Serial No.
Oscilloscope		
Signal or function generator		
Inductance meter or bridge		

TEXT REFERENCE

Section 23.9 LOOSELY COUPLED CIRCUITS
Section 23.10 MAGNETICALLY COUPLED CIRCUITS WITH SINUSOIDAL EXCITATION
Section 23.11 COUPLED IMPEDANCE

DISCUSSION

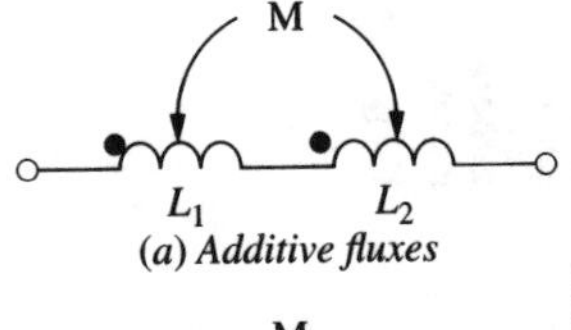

(a) Additive fluxes

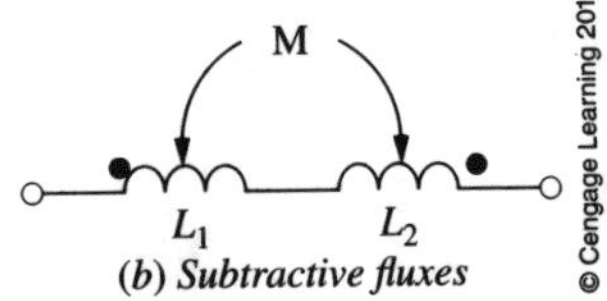

(b) Subtractive fluxes

FIGURE 28-1 Coupled coils.

If two coils with mutual coupling are connected in series so that their fluxes add as in Figure 28-1(a), their total inductance is

$$L_T^+ = L_1 + L_2 + 2M \tag{28-1}$$

where L_1 and L_2 are the self-inductances of coils *1* and *2* and M is the mutual coupling between them. If the coils are connected so that their fluxes subtract as in Figure 28-1(b), total inductance is

$$L_T^- = L_1 + L_2 - 2M \tag{28-2}$$

The mutual inductance between the coils can be found from the formula

$$M = (L_T^+ - L_T^-)/4 \tag{28-3}$$

(Since inductances L_1, L_2, L_T^+, and L_T^- can be measured with a standard inductance meter, you can use Equation 28-3 to determine mutual inductance experimentally.) Self and mutual inductances are related by the formula

$$M = k\sqrt{L_1 L_2} \tag{28-4}$$

where k is called the *coefficient of coupling*. For tightly coupled coils, $k = 1$; for loosely coupled coils, k is much less than 1.

Equations for Coupled Circuits. If coupled coils are connected as in Figure 28-2(a), their voltages and currents are related by the formulas

$$v_1 = R_1 i_1 + L_1\frac{di_1}{dt} \pm M\frac{di_2}{dt} \tag{28-5}$$

$$v_2 = \pm M\frac{di_1}{dt} + R_2 i_2 + L_2\frac{di_2}{dt} \tag{28-6}$$

where the sign of M is determined by the dot convention—that is, if both currents enter (or leave) dotted terminals, then the sign to use with M is positive, while if one current enters a dotted terminal and the other leaves, the sign to use is negative. For sinusoidal excitation as shown in Figure 28-2(b), Equations 28-5 and 28-6 become

$$\mathbf{V}_1 = (R_1 + j\omega L_1)\mathbf{I}_1 \pm j\omega M\mathbf{I}_2 \tag{28-7}$$

$$\mathbf{V}_2 = \pm j\omega M\mathbf{I}_1 + (R_2 + j\omega L_2)\mathbf{I}_2 \tag{28-8}$$

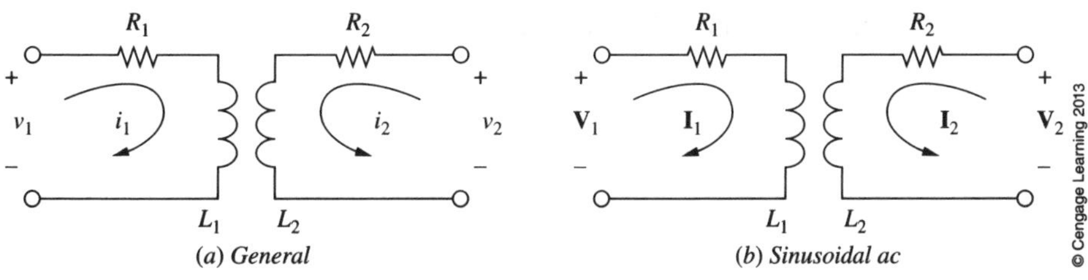

FIGURE 28-2 Coupled coils.

MEASUREMENTS

PART A: Coil Parameters and Mutual Inductance

1. a. Measure the primary and secondary inductances and resistances using an *LRC* meter or impedance bridge.

   L_1 = ___________ R_1 = ___________

   L_2 = ___________ R_2 = ___________

   b. Mark the coil ends for identification, then connect the primary and secondary coils in series as in Figure 28-1. Measure the inductance of the series connection, then reverse the connections and measure again.

   L_T^+ = ___________ L_T^- = ___________

   Use Equation 28-3 to compute M. M = ___________

   Use Equation 28-4 to compute k. k = ___________

   Is this transformer loosely coupled or tightly coupled? _______

   c. Use the results of Test 1(b) to determine the dotted ends for the coils. (Mark the dotted ends because you will need this information later.) Describe how you arrived at this conclusion.

   ____

   ____

PART B: Mutually Induced Voltage

2. Measure the value of the 10-Ω sensing resistor R_S, then assemble the circuit as in Figure 28-3.

   R_S = __________

Notes

You may have to choose different source values, depending on the transformer that you are using.

   a. Set the source voltage to 4 V_p (8 $V_{p\text{-}p}$) at $f = 1$ kHz—see Notes. Measure $\mathbf{V}_S$, magnitude, and angle (using the oscilloscope) and record below. Compute input current $\mathbf{I}_1$, magnitude, and angle. (You can leave the results in peak volts and amps, rather than convert to rms.)

   $\mathbf{V}_S$ = ________ ∠ $\mathbf{I}_1$ = ________ ∠

   b. With Probe 1 still measuring e, move Probe 2 to measure secondary voltage v_2. Measure magnitude and angle.

   $\mathbf{V}_2$ = ________ ∠

   c. Using the measured values of E, R_1, R_S, and L_1, compute current $\mathbf{I}_1$. Thus,

$$\mathbf{I}_1 = \frac{E}{R_1 + R_S + j\omega L_1}$$

   Compare to the result measured in (a).

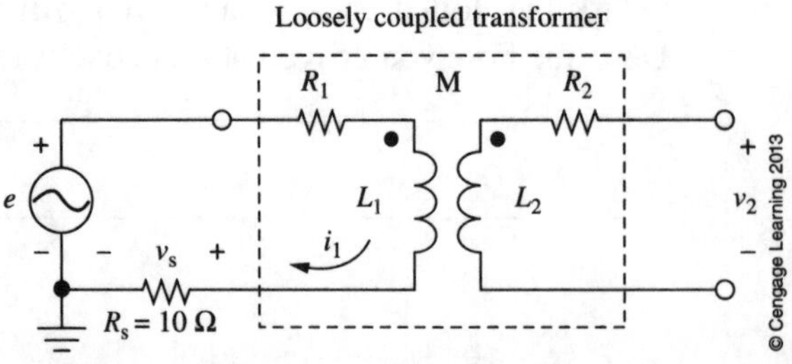

FIGURE 28-3 Circuit for Test 2. The secondary is open circuited here.

d. Compute $\mathbf{V}_2$ from the relationship

$\mathbf{V}_2 = j\omega M\mathbf{I}_1 =$

Compare to the result measured in (b).

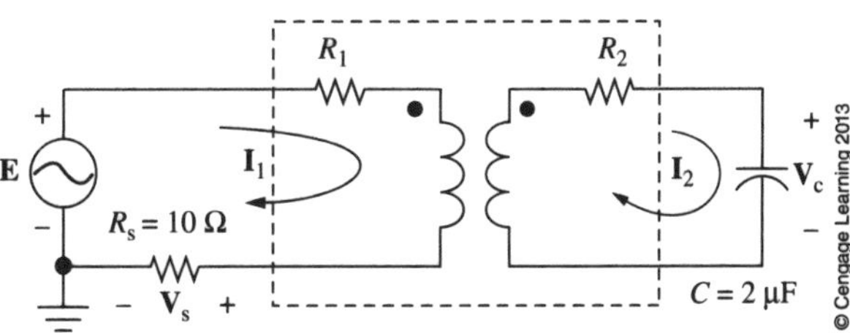

FIGURE 28-4 Circuit for Test 3.

PART C: Coupled Circuit Equations

3. Add 2 μF of capacitance to the secondary circuit as in Figure 28-4.
   a. Set e to 4 V_p (8 $V_{p\text{-}p}$). Carefully measure $\mathbf{V}_S$ and $\mathbf{V}_2$ as before.

   $\mathbf{V}_S =$ ________ ∠ $\mathbf{V}_2 =$ ________ ∠

   Compute current $\mathbf{I}_1 = \mathbf{V}_S/R_1 =$ ________ ∠

   b. Write mesh equations for this circuit and solve for $\mathbf{I}_1$ and $\mathbf{I}_2$. Using the computed value of $\mathbf{I}_2$, calculate $\mathbf{V}_2$. Compare $\mathbf{I}_1$ and $\mathbf{V}_2$ to the values measured in 3(a). How well do they agree?
   c. Solve for $\mathbf{I}_1$ using the coupled impedance equation

$$\mathbf{Z}_{\text{in}} = \mathbf{Z}_1 + \frac{(\omega M)^2}{\mathbf{Z}_2 + \mathbf{Z}_L}$$

   Compare to the measured value.

COMPUTER ANALYSIS

Use Multisim 11 (or later) or PSpice for the following.

4. Consider Figure 28-3. Using the parameter values that you measured in Part A, simulate this circuit and determine $\mathbf{V}_2$. Compare to the measured value.

5. Simulate the circuit of Figure 28-4. Using the parameter values that you measured in Part A, determine $\mathbf{I}_2$ and $\mathbf{V}_2$. Compare to the measured values.

PROBLEM

6. Write equations for the circuit of Figure 28-5 and solve for $\mathbf{I}_1$, $\mathbf{I}_2$, and $\mathbf{V}_2$.

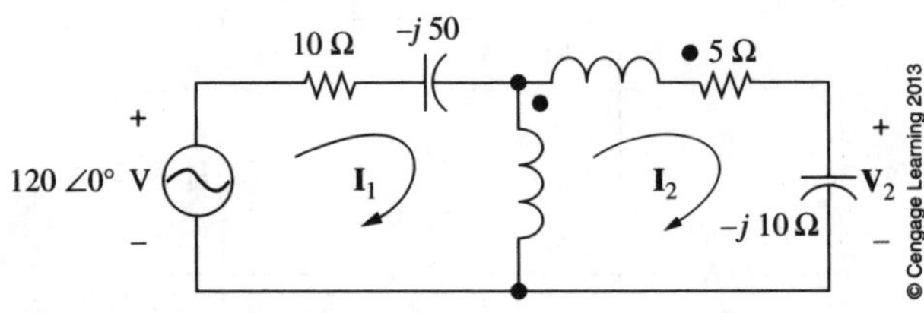

FIGURE 28-5 $X_{L_1} = 25\ \Omega$ $X_{L_2} = 30\ \Omega$ $X_M = 2\ \Omega$.

FOR FURTHER INVESTIGATION

Simulate the circuit of Figure 28-5 at $f = 60$ Hz.

APPENDIX A Resistor Color Codes

Figure A-1 shows a typical color-coded resistor together with the standard colors and corresponding designations. Table A-1 gives nominal resistor values for resistors having 5%, 10%, and 20% values.

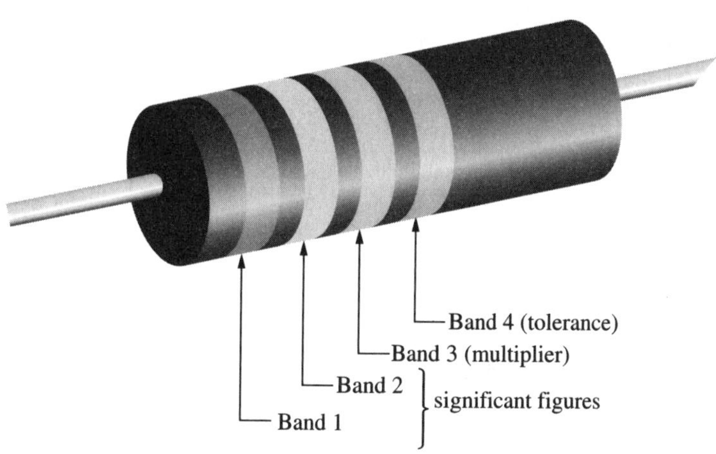

Resistor Color Codes—see Notes

Color	Band 1 Sig. Fig.	Band 2 Sig. Fig.	Band 3 Multiplier	Band 4 Tolerance
Black		0	$10^0 = 1$	
Brown	1	1	$10^1 = 10$	
Red	2	2	$10^2 = 100$	
Orange	3	3	$10^3 = 1\ 000$	
Yellow	4	4	$10^4 = 10\ 000$	
Green	5	5	$10^5 = 100\ 000$	
Blue	6	6	$10^6 = 1\ 000\ 000$	
Violet	7	7	$10^7 = 10\ 000\ 000$	
Gray	8	8		
White	9	9		
Gold			0.1	5%
Silver			0.01	10%
No color				20%

Notes

1. Resistors usually use four color-coded bands as shown.
2. Occasionally (for precision resistors, for example), a fifth band is used to denote reliability—see Chapter 3, Section 3.6 of the textbook (where we show a five-band variant of the color-code scheme).
3. Handy color-code converter calculators are available on the Internet. Search for *resistor color code calculator*.

FIGURE A-1 Standard resistor values.

TABLE A-1 Standard Resistor Values (5%—see Notes)

1.0	**10**	**100**	**1.0k**	**10k**	**100k**	**1.0M**	**10M**
1.1	11	110	1.1k	11k	110k	1.1M	11M
1.2	**12**	**120**	**1.2k**	**12k**	**120k**	**1.2M**	**12M**
1.3	13	130	1.3k	13k	130k	1.3M	13M
1.5	**15**	**150**	**1.5k**	**15k**	**150k**	**1.5M**	**15M**
1.6	16	160	1.6k	16k	160k	1.6M	16M
1.8	**18**	**180**	**1.8k**	**18k**	**180k**	**1.8M**	**18M**
2.0	20	200	2.0k	20k	200k	2.0M	20M
2.2	**22**	**220**	**2.2k**	**22k**	**220k**	**2.2M**	**22M**
2.4	24	240	2.4k	24k	240k	2.4M	
2.7	**27**	**270**	**2.7k**	**27k**	**270k**	**2.7M**	
3.0	30	300	3.0k	30k	300k	3.0M	
3.3	**33**	**330**	**3.3k**	**33k**	**330k**	**3.3M**	
3.6	36	360	3.6k	36k	360k	3.6M	
3.9	**39**	**390**	**3.9k**	**39k**	**390k**	**3.9M**	
4.3	43	430	4.3k	43k	430k	4.3M	
4.7	**47**	**470**	**4.7k**	**47k**	**470k**	**4.7M**	
5.1	51	510	5.1k	51k	510k	5.1M	
5.6	**56**	**560**	**5.6k**	**56k**	**560k**	**5.6M**	
6.2	62	620	6.2k	62k	620k	6.2M	
6.8	**68**	**680**	**6.8k**	**68k**	**680k**	**6.8M**	
7.5	75	750	7.5k	75k	750k	7.5M	
8.2	**82**	**820**	**8.2k**	**82k**	**820k**	**8.2M**	
9.1	91	910	9.1k	91k	910k	9.1M	

Notes

1. Resistors are available commercially in specific standard values with specified tolerances. Typical tolerances are 1%, 2%, 5%, 10%, and 20%.
2. Standard values are the same for each tolerance group, although not all values are available in each tolerance group—for example, all values shown in the above table are available in the 5% tolerance group, but only the boldfaced values are available in the 10% tolerance group.
3. Each tolerance group has an EIA classification. The 10% series is known as the E12 series, the 5% series is known as the E24 series, etc.
4. More detailed information may be found on the Internet. Search for *resistor standard values*.

Equipment and Components

STANDARD EQUIPMENT FOR BASIC LABS

1—Dual-channel oscilloscope
2—Digital multimeter (DMM) (Occasionally more are required.)
1—Power supply, variable, regulated
1—Signal generator (A function generator is required for a few labs.)
1—*LRC* meter or impedance bridge

SPECIALIZED EQUIPMENT FOR THE POWER LABS

2—Single-phase wattmeters (approximately 300-W full scale)
Three-phase ac source, 120/208 V

RESISTORS

Most resistor tolerances are not critical. However, unless otherwise noted, we recommend 5 percent tolerance resistors or better.

RESISTORS (1/4 W)

4.7 Ω, 6.8 Ω, 10 Ω, 15 Ω, 47 Ω, 75 Ω, 82 Ω, 100 Ω, 150 Ω, 180 Ω, 220 Ω, 270 Ω, 330 Ω, 470 Ω, 510 Ω, 680 Ω, 820 Ω, 1 kΩ, 1.2 kΩ, 1.5 kΩ, 2 kΩ, 2.2 kΩ, 2.7 kΩ, 3.3 kΩ, 3.9 kΩ, 4.7 kΩ, 5.1 kΩ, 5.6 kΩ, 6.8 kΩ, 7.5 kΩ, 9.1 kΩ, 10 kΩ, 12 kΩ, 18 kΩ, 20 kΩ, 39 kΩ, 47 kΩ, 51 kΩ, 75 kΩ, 100 kΩ, 180 kΩ, 330 kΩ, 3.3 MΩ, 5.6 MΩ, 10 MΩ

RESISTORS (OTHER)

(1/8 W) 470 Ω
(1/2 W) 47 Ω, 82 Ω, 470 Ω, 560 Ω
(1 W) 100 Ω, 220 Ω, 470 Ω
(2 W) 47 Ω, 75 Ω, 82 Ω, 100 Ω, 120 Ω, 270 Ω, 470 Ω, 510 Ω, 1 kΩ
(10 W) 10 Ω, 51 Ω
(25 W) 10 Ω
(200 W) 100 Ω (One for Lab 12. If you do the three-phase labs, you will need three.)
(200 W) 250 Ω (Three. Required for the three-phase labs only.)

CAPACITORS

4.7 nF, 2200 pF, 3300 pF, 0.01 μF, 0.022 μF, 0.047 μF, 0.1 μF, 0.22 μF, 0.33 μF, 0.47 μF, 1.0 μF

CAPACITORS (OTHER)

10 μF, 220 μF, 470 μF, electrolytic
30 μF, nonelectrolytic, rated for 120 VAC operation (One for Lab 12. If you do the three-phase labs, you will need three.)

INDUCTORS

1 mH. Powdered iron-core inductor (Hammond #1534A or equivalent)
2.4 mH (two). Powdered iron-core inductor (Hammond #1534C or equivalent. You need an inductor with low resistance to approximate the behavior of an ideal inductor.)
0.2 H (approximate value) inductor. Rated to handle 2 amps.
1.5 H iron-core inductor. (Approximate value. Value not critical.)

MISCELLANEOUS

Thermistor, 10 kΩ @ 25°C
Diode: 1N4004 or equivalent
Potentiometers: 5 kΩ, 10 kΩ, 20 kΩ
Transformers: 120/12.6 V (filament)

Using Microsoft Excel as a Graphing Tool

Graphs are an integral part of examining and analyzing the characteristics of electrical and electronic circuits. While graphs have traditionally been sketched using paper and pencil, there are now numerous software applications that provide an alternative to hand drawing. In this section, we use Microsoft Excel™ to demonstrate how to obtain graphical outputs that will help you to visualize circuit operation. You will learn how to convert a table of data into meaningful graphs, complete with table headings, axis labels, and suitable output displays. You will also learn how to use logarithmic scales to obtain graphs of filter circuits.

EXAMPLE C-1

Use Excel to obtain a sketch of power P_L versus resistance R_L for the simple series circuit shown in Figure C-1.

FIGURE C-1 Series circuit with a variable resistor at the output.

Solution

Let us assume that you constructed the circuit in Figure C-1 in the laboratory and that you varied the load resistor R_L in increments of 1 Ω to obtain the following measurements and calculations:

TABLE C-1 Voltage, Current, and Power Measurements for a Series Circuit

R_L (Ω)	V_L (Ω)	I_L (A)	P_L (W)
0	0	2.000	0
1	1.667	1.667	2.778
2	2.857	1.429	4.082
3	3.750	1.250	4.688
4	4.444	1.111	4.938
5	5.000	1.000	5.000
6	5.455	0.909	4.959
7	5.833	0.833	4.861
8	6.154	0.769	4.734
9	6.429	0.714	4.592
10	6.667	0.0667	4.444

The data from the lab measurements will first need to be entered into the Excel spreadsheet as shown in Figure C-2. Once this is done, the mouse is used to select the data that is to be graphed. Next, we need to select the type of chart by left-clicking on the *Insert* tab, selecting *Scatter* charts, and then opting for the *Scatter with Smooth Lines and Markers*. These steps are all illustrated in Figure C-2.

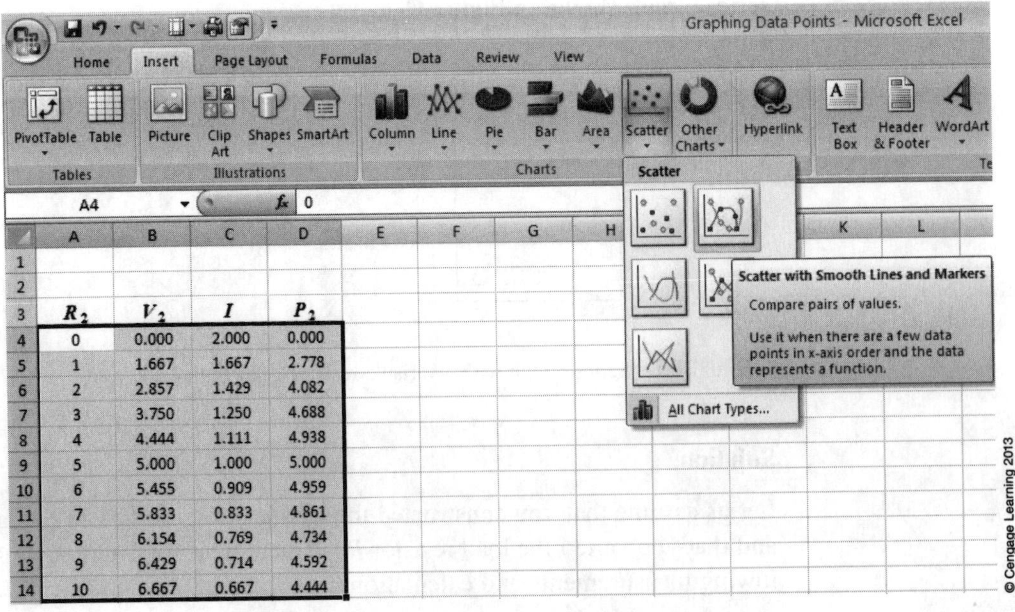

FIGURE C-2 Excel spreadsheet display for a simple series circuit.

By following the previous steps, Excel places the graph shown in Figure C-3 into the spreadsheet.

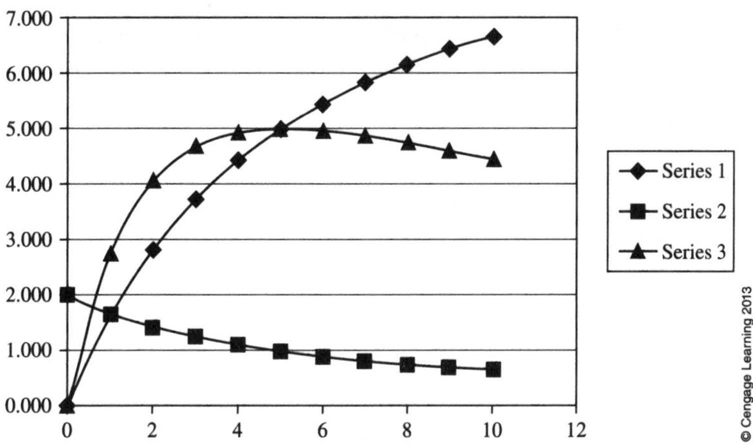

FIGURE C-3 Graph of voltage, current, and power.

While the data shown in Figure C-3 are interesting, they do not give an observer much useful information about the contents. The graph can be made more useful by removing the unnecessary data plots and by labeling the abscissa (X-axis) and ordinate (Y-axis) with appropriate units.

Since we would like to view only the graph of power P_L as a function of resistance value R_L, which is shown in Figure C-3 as Series 3, we will need to remove the other series of data points. This is done by right-clicking anywhere in the chart area to bring up the pop-up menu shown in Figure C-4.

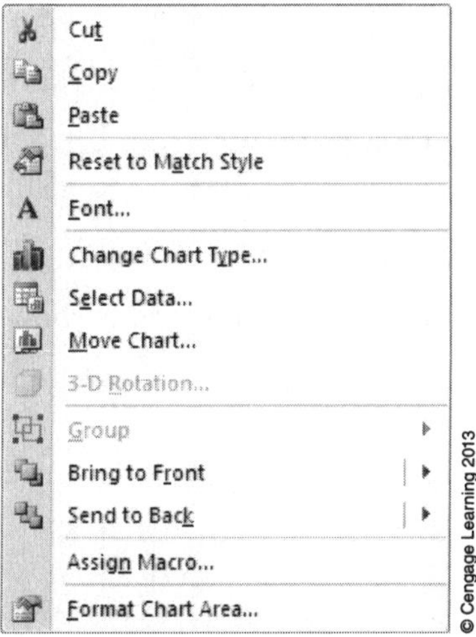

FIGURE C-4 Chart pop-up menu.

Next click on *Select Data...* to bring up the *Select Data Source* box shown in Figure C-5.

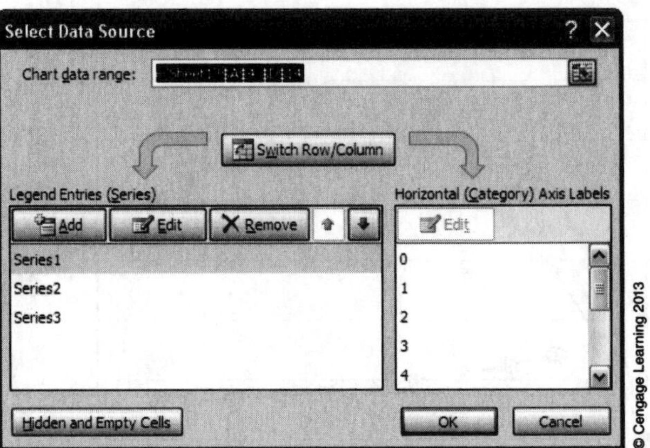

FIGURE C-5 Select Data Source box.

It is here that we remove Series 1 and Series 2 data points from the graph by selecting these series and then clicking on *Remove*. Finally, the chart is labeled and formatted by clicking anywhere in the chart area to bring up the Chart Tools as shown in Figure C-6. The chart title, axis titles, and axes are changed by selecting the appropriate buttons and then making the desired changes under the *Layout* tab.

FIGURE C-6 Chart Tools.

The final result of the graph is shown in Figure C-7.

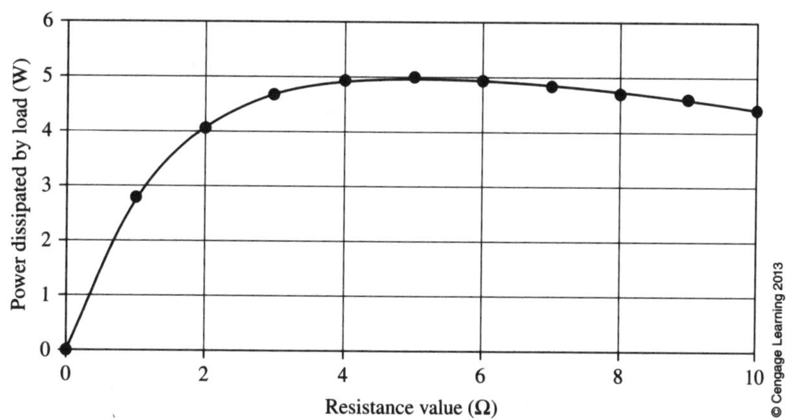

FIGURE C-7 Graph of power as a function of resistance value.

PRACTICE PROBLEM C-1

Use the method shown in Example C-1 to obtain graphs of:

a. Voltage V_L as a function of resistance value R_L and
b. Current I_L as a function of resistance value R_L.

The graph shown in Figure C-7 has the axes shown with "linear scaling," which means that each horizontal division has the same value, in this case 2 Ω. While for many graphs, this is a useful scaling factor, this is not always the case. The following example shows that linear scaling does not result in the most useful graph. Instead, we will find that it is more useful to use logarithmic scaling.

EXAMPLE C-2

Use Excel to obtain a sketch of voltage gain A_V(dB) versus frequency ω (rad/s) for the circuit shown in Figure C-8.

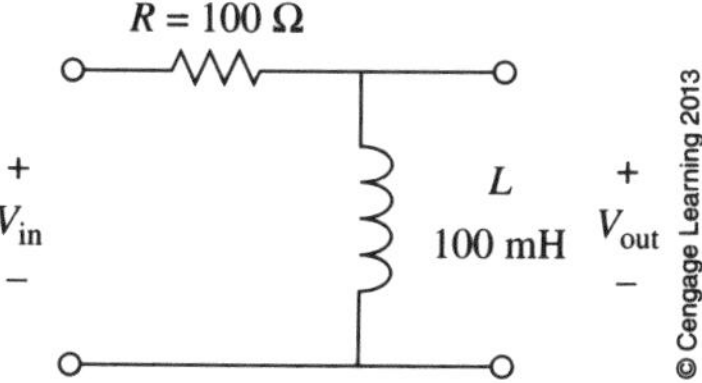

FIGURE C-8 *R-L* high-pass circuit.

Solution

Let us assume that you have built the previous *R-L* high-pass circuit and obtained the measurement results shown in the Excel spreadsheet of Figure C-9.

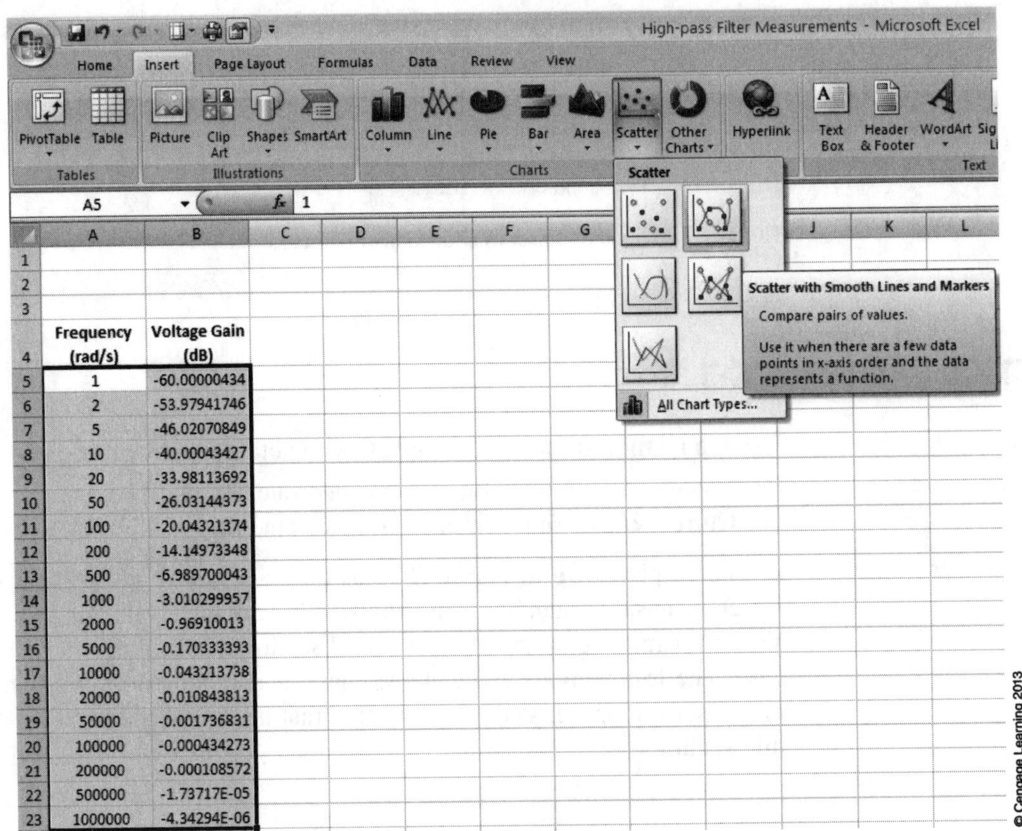

Row	Frequency (rad/s)	Voltage Gain (dB)
5	1	-60.00000434
6	2	-53.97941746
7	5	-46.02070849
8	10	-40.00043427
9	20	-33.98113692
10	50	-26.03144373
11	100	-20.04321374
12	200	-14.14973348
13	500	-6.989700043
14	1000	-3.010299957
15	2000	-0.96910013
16	5000	-0.170333393
17	10000	-0.043213738
18	20000	-0.010843813
19	50000	-0.001736831
20	100000	-0.000434273
21	200000	-0.000108572
22	500000	-1.73717E-05
23	1000000	-4.34294E-06

FIGURE C-9 Excel spreadsheet display for an *R-L* high-pass filter circuit.

As in Example C-1, we need to select the data and the type of chart. As before, we select a *Scatter Chart with Smooth Lines and Markers*. The graph that we observe will appear as shown in Figure C-10.

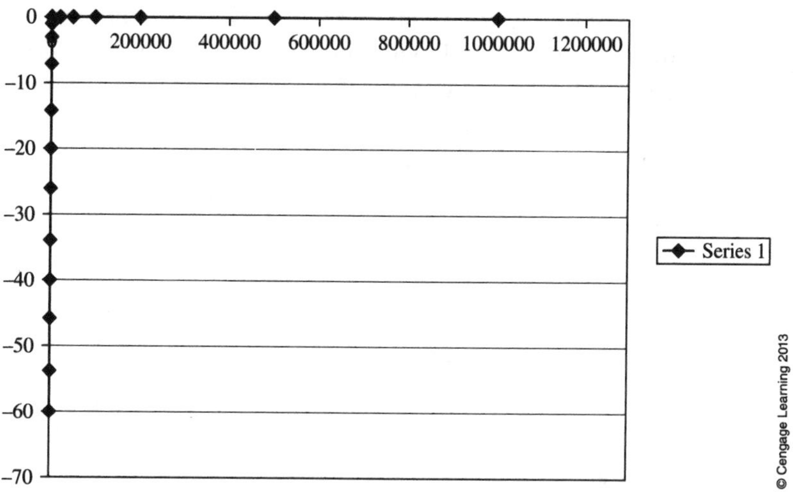

FIGURE C-10 Voltage gain versus frequency on linear abscissa.

Figure C-10 is not very useful since it appears that the graph consists of two straight lines that appear to intersect at a frequency of 0 rad/s. By making a few changes to the graph, we will obtain a graph that is much more useful.

First, we click anywhere in the chart area to bring up the Chart Tools with the *Design, Layout,* and *Format* tabs. Then, since we wish to change the layout of the graph, we select the *Layout* tab. In order to change the horizontal axis settings, we next click on the buttons to bring up the options shown in Figure C-11.

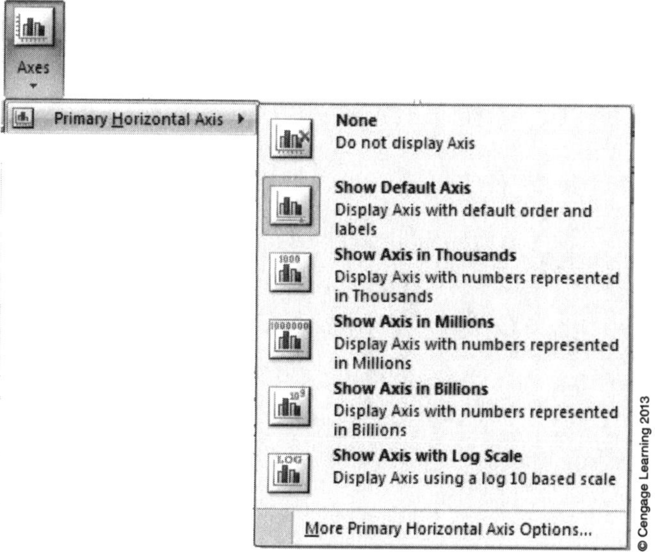

FIGURE C-11 Horizontal axis settings.

By selecting the *Show Axis with Log Scale* button, we end up with a much more useful graph. Finally, by using other Layout options, we have a graph complete with a title, axis labels, and grid lines as shown in Figure C-12. In order to show the axis label at the bottom of the graph, click anywhere on the horizontal axis and then select *Low* from the options under the *Axis Labels*. From this graph, we see that the cutoff frequency of the *R-L* filter occurs at a frequency of 1000 rad/s.

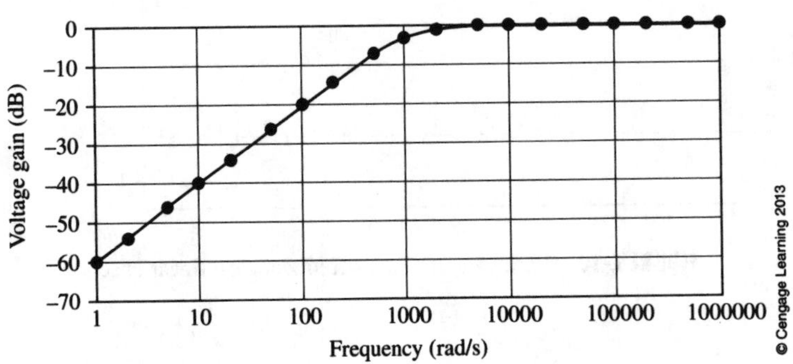

FIGURE C-12 Voltage gain versus frequency on logarithmic abscissa.

NOTES

NOTES

NOTES

NOTES

NOTES

NOTES

NOTES

NOTES

NOTES

NOTES